Trace Your Roots with DNA

Using Genetic Tests to Explore Your Family Tree

Notice
This book is intended as a reference volume only. Mention of specific companies, organizations, or authorities in this book does not imply endorsement by the publisher, nor does mention of specific companies, organizations, or authorities imply that they endorse this book.

Internet addresses and telephone numbers given in this book were accurate at the time it went to press.

Printed in the United States of America
Rodale Inc. makes every effort to use acid-free ♾, recycled paper ♻.

"Sample Y-DNA Haplogroup Descriptions" is used by permission of Family Tree DNA and The Genomics and Technology Core at the University of Arizona.

"The 7 Daughters of Eve Descriptions" is used with permission of Dr. Bryan Sykes and Oxford Ancestors.

"mtDNA Migration Map" is used by permission of MITOMAP: A Human Mitochondrial Genome Database. <http://www.mitomap.org>, 2004.

Photograph by Kathy Peacock

Book design by Gavin Robinson

Library of Congress Cataloging-in-Publication Data

Smolenyak, Megan.
Trace your roots with DNA : using genetic tests to explore your family tree / Megan Smolenyak Smolenyak and Ann Turner.
p. cm.
Includes index.
ISBN 1-59486-006-8 paperback
1. Genealogy. 2. DNA—Genealogy—Handbooks, manuals, etc. 3. Genetics—Genealogy—Handbooks, manuals, etc. I. Turner, Ann. II. Title.
CS21.S58 2004
929'.1'072—dc22 2004014980

Distributed to the trade by Holtzbrinck Publishers

2 4 6 8 10 9 7 5 3 1 paperback

For Seton Shields—
Your mtDNA was the least of the gifts you've given me!
—MSS

For Jim Turner—
Whose gift of a DNACUZN license plate symbolizes his support
—APT

Contents

Acknowledgments

Every time a book finds its way to your local bookstore or library, one or two names appear on the cover. The authors hog the credit, but the reality is that it takes a significant team effort to bring a book to life. This book is no exception—and though we live in fear of neglecting to mention someone (and fervently beg your forgiveness, if this has happened!)—we'd like to recognize those who have in some way co-authored this book with us.

Linda Konner, our agent, once again found just the right home for our fledging book-to-be and handled all aspects with her usual ease. Mariska van Aalst and Amy Super of Rodale championed our vision (and sometimes took the blinders off our eyes when we overlooked the obvious) while Emily Williams, Jessica Roth, Rose Panetta, Chris Rhoads, Gavin Robinson, and countless others labored to bring the separate pieces together at last.

Many experts in the field of genetealogy kindly shared their wisdom with us and endured repeated rounds of questions and clarifications. We are especially grateful to those who were willing to openly speculate on the future and trust us with information that was confidential at the time of the writing of this book. We thank Terry Carmichael, Tony Frudakis, Alastair Greenshields, Bennett Greenspan, Ripan Malhi, Terry Melton, Ugo Perego, Diahan Southard, Richard Villems, and Bruce Walsh. Authors and scientists who educated us and inspired us to make the leap from theory to practice include Scott Woodward, Michael Hammer, Luigi Cavalli-Sforza, Mary-Claire King, Steve Olson, Stephen Oppenheimer, John Relethford, Bryan Sykes, and Spencer Wells.

As is evident throughout this book, we also benefited from the expertise of other clusters of experts, such as professional genealogists and pioneering genetealogists, who have figured things out and developed resources that help the rest of us shorten our learning curves. Our gratitude goes to Joe Beine, John Chandler, Bob Dorsey, Kevin Duerinck, David Faux, Linda Hammer, Nel Hatcher, Bill Hurst, Brent Kennedy, Charles Kerchner, Louis Loccisano, Ana C. Oquendo Pabón, José Antonio Oquendo Pabón, Tom Osborne, Chris Pomery, David Roper, Christine Rose, Bonnie Schrack, and all those who enlighten us on the Genealogy-DNA Mailing List. Thanks also to Carey Bracewell, Everett Christmas, Don Green, Tara Robinson and Suzanne Walker for their much appreciated contributions.

And as if project managers don't have enough on their hands administering their DNA studies, many of them generously gave of their time to respond to a detailed survey and additional communications. We are deeply indebted to all of the following for sharing their experiences and insights, and allowing us to tell their stories: Terry Barton, Bill Bailey, Wayne Bates, Eddie Bennett, John A. Blair, Gary Blakely, Gregg Bonner, Georgia Kinney Bopp, Larry Bowling, David W. Brown, David Clifford, Barry Collett, Pieter J. Cramwinckel, Nancy Custer, Jay Dixit, John German, Wade Glascock, Clarke Glennon, Daniel Guggisberg, D. Harper, Phillip Hawkins, Kenneth R. Herrick, Ray Hill, Geoffrey Hodgson, Mary Lou Hudson, James Reynolds Hull, Roy Hutchinson, Melissa Jones, Norman Jordan, Peter A. Kincaid, Hikaru Kitabayashi, Steve Laymon, Ken Lennan, Barbara McCarthy, Cliff McCarthy, Bill McCeney, Janice McGough, Richard McGregor, Sharon S. Miller, Ferd Mireault, Jesse Moore, Steven C. Perkins, Ken Rockwell, Mary Harkey Russell, Michael Rutledge, Thomas J. Schmidt, Sharron Spencer, Justin (Howery) Swanstrom, Lannie G. Walker, Sr., and Dan Wharton. Thank you for your enthusiasm, sense of adventure, hard work, and most of all, for introducing so many to the world of genetealogy!

Megan

Brian Smolenyak insisted that this book be born, knowing full well the "Sarah Heartburn" moments that would ensue, and calmly weathered yet another gestation period with constant support. Stacy Neuberger did the impossible repeatedly with boundless supplies of encouragement and nary a word of grumbling. Thanks, SAS! Seton Shields and Ray Freson sacrificed a healthy chunk of their vacations wading through an intense early draft to supply much needed feedback, as well as spot-on recommendations for improvement. George C. Smolenyak also passed on some holiday festivities for our benefit and gave me an entertaining genetic heritage that furnished several of the examples in this book! Laura Tinsley provided a fresh, scientific perspective that proved invaluable, and once again, Anna Grace Harding provided perspective, period! And finally, my coauthor Ann Turner furnished in-depth expertise that would have taken me decades to acquire. Thanks, Ann, for the intensive education!

Ann

I owe a debt of gratitude to all my ancestors (who somehow managed to survive a perilous journey down through time), many collateral relatives (who donated some DNA to satisfy my curiosity), my sister, Mary Evans (who shares my enthusiasms and some—but not all—of my genetic traits), my husband, Jim Turner (who tapes cartoons to my computer monitor), and my coauthor Megan Smolenyak (who turned a mere velleity into reality).

Introduction

Welcome to the World of Genetic Genealogy

It's probably happened to you. Maybe you were at a meeting, and an acquaintance introduced you to his coworker who coincidentally had the same surname as you. Back in your school days, you might have found yourself seated next to another kid with the same last name. Perhaps you took a little extra pride in that Olympic champion whose surname was identical to yours. Or maybe you were channel surfing and did a double take when the name under one of those talking-head experts happened to match your own. And when it happened, you probably wondered, "Are we—could we be—related?"

At some level, we all have a desire to know about our origins, and this name curiosity is just one symptom. Some of us join the quest early in life; others are successful at ignoring the pull for many years, but sooner or later, it gets us. This is what Alex Haley was speaking of when he said, "In all of us, there is a hunger, bone-marrow deep, to know our heritage—to know who we are and where we have come from. Without this enriching knowledge, there is a hollow yearning. No matter what our attainments in life, there is still a vacuum, and emptiness, and the most disquieting loneliness."

Genealogy offers a means to satisfy this hunger, and you may well be one of the more than 40 million roots-addicted in the United States alone. The two of us writing this book certainly count ourselves among that number and are delighted with the millions of new family history playmates the recent years have brought.

Revolutionary Times

But the past decade has gifted us with more than fellow ancestry detectives; it's given us incredible new tools, the most obvious being the Internet. If you can remember that long ago—back to that distant time when you took your timid baby steps on the Internet—what did you do the first time you encountered a search engine? Unless you've got a total lack of self-interest or more discipline than 99 percent of us can muster, you typed your name. You did a vanity search, didn't you?

Whether you recognized it or not, you were answering that bone-marrow-deep call and doing basic genealogy. The Internet has revolutionized our world by making it exponentially easier to find not only useful resources but also our second, third, and fourth cousins. And vanity searches have often been the first step in the process of answering that age-old question: *Where do I come from?*

This book is about the next revolution. Genetic genealogy—DNA testing done with the aim of learning about one's heritage—provides a key for unlocking some secrets that the paper trail can never reveal—that would otherwise be unknowable. If the Internet made it possible to find our second and third cousins, *genetealogy* (ge-neh-tee-ol-o-gee)—the word we've coined for the merger of genetics and genealogy—will make it possible to find our twelfth and sixteenth cousins.

If you've ever wondered if you're related to someone with the same surname, been curious about your African ancestry, or tried to figure out if your grandfather was really adopted, genetic genealogy can answer these questions—and more.

Genetics for Genealogists

By now, you're probably wondering, "What's this all about?" You're probably a little bit curious, and you may be more than a little bit skeptical about the benefits of reading this book.

Let's be clear up-front that this book is about *genetics for genealogists,* not vice versa. We'll explain what genetic testing is and how it's carried out through DNA research projects. Whoa, you say! What's wrong with my DNA that it needs to get tested? And while I may be the family sleuth, I'm not about to get involved in starting some kind of scientific research program.

First of all, the term "DNA testing" is a somewhat vague or imprecise way of labeling this process. Perhaps "DNA typing" or "genetic charting" would be more to your liking, but DNA testing is what it's most commonly called. However you refer to it, though, it is simply the process of determining whether two individuals share a common ancestor by comparing an infinitesimal fragment of their respective DNA. (There are a couple of other, more generic tests, but we'll get into all that a little later.) It's easy, and nobody has to go to the hospital lab or give blood. And as you'll learn, these fragments do not constitute a chart of your medical makeup, compromise your privacy, or qualify you for a new file in the FBI archives!

A DNA "project" is the systematic collection of this data for a population of related individuals. It's not a formal project—there's no staff or outside entity looking over your shoulder. Usually, it's a voluntary exercise that isn't much different than contacting existing or suspected relatives by letter or e-mail to fill in some of the blanks in your family tree.

It is, however, an amazing new resource for climbing your family tree, and whether or not you're anxious to jump right in with your own project, by the time you finish reading this book, you'll be knowledgeable about what DNA testing is and isn't, and what it can and can't do. In a sense, you'll become the family expert on this burgeoning new research tool. And if someone else running their own project contacts you, you'll be in a much better position to evaluate whether or not to participate and how it might fit in with your own sleuthing efforts.

In order to understand the process, we're going to delve into the science of genetics, but we promise you can absorb as little or as much of the technical aspects of this field as you like and still come away with a firm general understanding of what it is and how it might fit into your or some other family group's fact gathering. If you choose to, you can be the first at your family reunions to toss around terms like Y-DNA, mtDNA, haplotype, and mutation. But even if you're not ready just yet to dash out and order up some testing, when the time comes or the need arises, you'll be well equipped to use this remarkable new tool.

The Secrets in Our DNA

Twenty years ago, DNA was a term we encountered only in our science textbooks, but now it's part of our everyday lives and vocabulary. Turn on your car radio, and hear the latest about cloning. Pick up the paper, and read about how it factors into the hunt for a cure for Parkinson's. Plop down at the end of the day for an episode of *CSI* or any of a handful of shows that center on DNA as a crime-solving tool. Most of us don't really understand it, but that doesn't stop us from being fascinated by it.

In the year 2000, a White House ceremony marked a milestone in the Human Genome Project, an outline listing the sequence of 3 billion DNA bases. It took 10 years, hundreds of millions of dollars, and the cooperative efforts of laboratories around the world to compile. At the time, many understood that this accomplishment could help us prevent diseases and solve crimes, but few realized its potential for telling us about our past. Scientists of various stripes soon recognized this possibility. Geneticists, anthropologists, and archaeologists traveled the globe, sampling the DNA of peoples both living and long dead.

What secrets did our DNA offer up? We were stunned to learn just how young our species is, that we are ultimately all

African, and that we are all cousins. (Incidentally, scientists have indeed determined that all of mankind originated in Africa, but some of us hung around there longer than others!) In fact, we are so closely related that 99.9 percent of our DNA is identical. Yet the remaining 1⁄10 of a percent, that one part in a thousand, translates into about 3 million differences between any two of us (with the exception of identical twins, of course). We are all alike, yet we are each unique. Curiously, most of these variations are found in "junk" DNA—sections of DNA that serve no apparent purpose, yet preserve our ancient history because they are copied more or less faithfully for generation after generation.

If we could track a little snippet of DNA that shows one of these variations, we'd follow a path that meandered back through the generations: It came from one of our two parents, one of our four grandparents, one of our eight great-grandparents. The numbers double with each generation: 16, then 32, 64, 128, 256. By the time we reach 10 generations, that little snippet of DNA could have come from any one of 1,024 potential ancestors. But if you pluck just any one of those ancestors from your family tree, you might not have any remnants of his DNA at all.

How then can we use DNA to trace our genealogy? We know that all of our DNA came from somewhere, but that's not much help if we can't put our finger on the right slot in our family tree. Fortunately, there are two kinds of DNA that follow a straight line instead of a meandering path—DNA found on the Y chromosome and mitochondrial DNA (mtDNA, pronounced em-tee-DNA). By a happy coincidence, the straight line for the Y chromosome is the same as the surname line in many cultures, and the straight line for mtDNA tracks the often-elusive female side. It's the possibilities offered by these two kinds of DNA that will be the focus of our attention for much of this book, although we won't neglect to explore what DNA can reveal about the rest of your family tree.

These DNA heirlooms passed to you not through the probated wills of your ancestors but by the sheer will of your ancestors to survive and pass on their legacy. Until recently, there was no way to take a peek at the inheritance hidden inside your cells, but the Human Genome Project developed methods that could be adopted by commercial laboratories. The very same year that the White House held that press conference, genealogical testing companies began offering DNA services to the general public for the first time.

All new technologies take awhile to catch on. Usually, they start with a handful of pioneers (so called early adopters) who gradually spread their enthusiasm to others. As their numbers increase—and prices inevitably come down—they begin to multiply more rapidly. Eventually, they hit critical mass or what Malcolm Gladwell recently described in *The Tipping Point*—"that magic moment when an idea, trend, or social behavior crosses a threshold, tips, and spreads like wildfire."

That's where we are with genetealogy. Two to 3 years ago, we were among the first kids in this particular playground, but now it's getting pretty crowded. And yet, the technology is young enough that a pioneering spirit and contagious sense of zeal are still very much in evidence. We invite you to join the other trailblazers who have married genetics and genealogy to discover their own past.

How to Use This Book

We hope you'll use this book as your coach for joining, launching, or furthering your own genetic testing project. And just as any decent coach knows, no two people require exactly the same assistance. For that reason, we've made this book as flexible as possible to accommodate varying levels of expertise and patience.

To make it easier to digest, we've used plenty of examples, many of which may sound vaguely familiar. If you've heard about the Romanovs, the Titanic baby, or almost any history mystery that's been tackled through the use of DNA, you'll read about it here. But you'll also be introduced to everyday folks—not scientists or millionaires—whose desire to learn about their roots has led them to be the explorers of the world of genetealogy. Incidentally, both of us fit that description, so we'll be sharing some of our own experiences as well. And in case you're curious, most of the names in the stories you'll read are real, but in a few instances, we have changed them at the request of those involved. When referring to DNA project managers—those folks who generously volunteer their time to oversee a particular DNA study, we've also included the surname that's their focus in parentheses, such as this: Mary Lou Hudson (Cox, Anderson). We've done this to maximize the chance that some of you will fortuitously stumble onto a mention of a project that's already being conducted on your own name.

Since both genealogy and genetics are loaded with jargon and can be somewhat intimidating, we'll start with brief introductions to both before explaining how the two can be used in tandem. From there, we'll move on to a series of four chapters, each focusing on different types of testing that are currently available. Because the most prevalent tests are conveniently also the easiest to master, we've arranged the chapters roughly in order of popularity and complexity, so that you'll read first about the tests you are most likely to take and have the opportunity to gradually build your overall understanding chapter by chapter. Although they're not strictly roots-oriented, we've also included a chapter on what we call close kin tests—paternity, siblingship, etc. We've done so because it may be useful if you've hit a brick wall early in your research (e.g., maybe Grandma was adopted), and as is apparent from all the DNA testing featured on talk shows, it's an area of curiosity to most of us. All of these chapters will be liberally sprinkled with real-world cases, so you'll be able to see a variety of possible applications.

A chapter on how to join a project or manage your own study will walk you through the basics as well as equip you to deal with potentially thorny issues such as cost and privacy, should you decide to launch your own project. Additional chapters cover tactics for locating participants (often strangers) with the "right" DNA to answer your questions, and reporting and interpretation resources to help you get the most out of your test results and share them with others.

If you're completely new to all of this, we strongly recommend that you read this book straight through, but if you're a family history pro, we suggest you skip to Chapter 2, Genetics Essentials. If you're one of those pioneers who's already running your own study, you might be tempted to leap all the way to Chapter 8, but we hope you'll take the time to browse Chapters 4 through 7 to learn more about what your peers are doing. Maybe you'll find some ideas worth borrowing. So feel free to absorb every word or dip in here and there—whatever works best for you!

Part 1

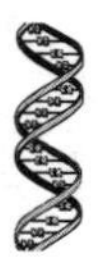

The Fundamentals

1

If You're New to Genealogy

Just by picking up this book, you've revealed that you're curious about your roots. And if you're curious about your roots, you're in good company! Millions of people are digging into the past, and the good news is that it's easier today than ever before. In fact, we tend to tell so-called newbies that they were smart to wait! One of us has been researching her family for 33 years and has learned more in the past 8 years than in the first quarter of a century.

This is largely because of the impressive and ever-growing collection of online and other resources. In fact, we recently conducted an experiment involving 33 popular genealogical resources. When we inspected the list, we discovered that 11 of them were not available a decade ago (such as www.ellisisland.org, www.findagrave.com, and the 1930 U.S. census released in 2002) and 22 of them existed, but were less accessible (such as resources now searchable at www.familysearch.org, the *Social Security Death Index*, which conveniently lists most Americans who have died since 1962, and every name indexes for the 1860, 1870, 1880, and 1930 U.S. census). Just 10 years ago, the notion of being able to search fully indexed and digitized records at home in your pj's was a wild fantasy. But it's reality now, and like good 21st-century citizens, we already take this previously unimaginable ability for granted!

If you're new to the game of family history research, we invite you to spend a little time with us as we cover the basics. You may be anxious to jump right into DNA testing, but a bit of genealogical effort invested upfront will ensure that you won't find yourself staring at a report with a bunch of numbers and scratching your head. Your venture into genetealogy will be much more fruitful if you learn some ABCs. Even if you're an old pro, you may wish to consider giving this chapter a skim to acquaint yourself with some genealogical nuances as they pertain to DNA testing.

Thousands of books and millions of Web sites are devoted to family history, so we won't go into great depth (although you'll be able to find more resources in Appendix A). Rather, we'll share enough to help you avoid the most common pitfalls that even seasoned roots-seekers occasionally fall prey to. Developing a few good habits early can save you days, weeks, and even months of frustration, so we'll start with some useful guidelines.

Do Your Homework

We know. You want to get a running start. You want to jump on the Internet or dash out to the nearest library or archive to find everything you can on your family. So don't hate us for telling you that you need to start at home.

Surfing the Internet is so easy—and on the surface—so gratifying, but it's apt to be a time-waster if you haven't done your groundwork. If you have a common name, you'll find yourself overwhelmed with the millions of sites that might shelter tidbits about your family. And even if your name is somewhat unusual, you'll probably be startled by how many hits you get when you type it in. A search on the borderline freakish name of Smolenyak will serve up almost 1,500 listings to wade through, so heaven help you if your name is Van Aalst (20,000+), Smithson (175,000+),

Pennington (1 million+), or Nichols (2.4 million+)! Maybe there aren't a lot of people with your name in your town, but there are a lot on the planet, and researching or contacting them all is an inefficient method of learning about the ones in your family tree.

And while conducting on-site research in records repositories should definitely be on your genealogical to-do list, it's best to look first for the treasures that may be lurking in your closets, drawers, basements, and attics—and especially the minds of your older relatives. Many a genealogist has been chagrined to finally discover an elusive maiden name, birth date, or village of origin after a year of research, only to find this same information in a suitcase of old papers tucked into the corner of their own cellar. And there's nothing quite as deflating as calling your great-aunt Mildred to announce your latest discovery only to have her reply, "I could have told you that."

To give you a feel of what you're looking for when you play detective in your own home (or maybe your parents', if they'll let you), here are a few items that are the equivalent of genealogical gold:

- Birth, marriage, and death certificates
- Newspaper clippings including obituaries and wedding and anniversary announcements
- Naturalization and citizenship papers, including passports and visas
- Religious records (baptismal, Bar Mitzvah)
- Family bible
- Letters and addressed envelopes
- Diaries and journals
- Photo albums
- Any other documents pertaining to your ancestors (military, school, occupational, business, land, legal)
- Heirlooms such as engraved items, samplers, and quilts

And if there's anyone in the family your age or older (and by family, we mean even that second cousin who lives in Denver whom you haven't seen since your sister's wedding back in 1984), pick up the phone! Not next week or next month—today! Talk with them immediately, if only to arrange a time to meet or call to learn more. Do not allow yourself to become one of the millions who bemoan the fact that they didn't ask questions when their parents/grandparents/aunts/uncles/cousins were still alive. If you were to eavesdrop at a genealogical conference or research venue, you'd be amazed how many times you'd hear comments starting with "If only I had . . ."

Draw up an initial list of questions you'd like answered, and be sure to ask about anyone else they think you should contact. Virtually every family has an avid genealogist, and you can be sure that older family members will know who that person is because they will have already spoken with him. Canvassing the relatives like this will turn up countless details that may not have trickled down your direct line. Your mother may not know that her grandmother had the maiden name of Doran, but there's a reasonable chance that one of her cousins does.

Don't Believe Everything You Hear or Read

Yes, we've just asked you to contact your assorted kinfolk and pummel them with questions, but that doesn't mean you have to accept everything they say as fact. We know this in everyday life, but for some reason, we seem to forget this when it comes to our family history. Because we obtain much of this information from relatives—and everyone knows that Great-aunt Tillie never lies—our family lore takes on the veneer of absolute truth. Many of us will accept oral tradition over the documented paper trail.

How many of us, for instance, have a family legend about our names being changed at Ellis Island? Great-grandpa couldn't speak English and had a difficult time communicating, so the inspectors listed him with the same name as the fellow in line before him. Or the immigration officials couldn't pronounce the surname, so they lopped off the last syllable or two. The reality is that the manifests were created overseas, and the officials here—assisted by translators who spoke all the languages of the immigrants—did their best to confirm the details. If a name was changed, it was almost always because the immigrant wanted it changed, and it usually happened *after* he walked the halls of Ellis Island. But try telling that to someone who heard the tale from Grandma.

No matter how sincere the intentions of the storyteller, chances are that a little distortion has crept in over the years. Through a combination of misunderstanding, forgetfulness, embellishment, and deliberate twisting, family lore morphs over the generations. Inevitably, there's a kernel of truth—and sometimes 95 percent will be accurate—but routinely accepting all family folklore as fact will usually throw off your research. Blinded by the tale, we get locked into a paradigm that prevents us from discovering the reality.

And don't think it's necessarily accurate just because it's in black and white. Documents contain errors too—plenty of them. For example, one of our grandfather's birth certificates listed Greece as the birthplace of his mother. After several years had been squandered trying to find the elusive Greek great-grandmother, it was discovered that she had emigrated from Poland. How could Greece and Poland be muddled? She was of the Greek Catholic faith, and her religion and nationality had been confused on the certificate.

This doesn't mean you should ignore the family stories and discount everything you find—or that you're excused from interviewing your relatives! But it does mean that you should examine every piece of information with a critical eye. Think of each family tale as a hypothesis you can prove or disprove through your research—maybe even through DNA testing.

Capture What You Learn

Now that you've learned all these wonderful details (not to mention that story you had never heard about the time your grandmother got angry at her sister and cut off only *one* of her pigtails) don't let it escape! Most of us know that our memories aren't foolproof, but we sometimes give ourselves too much credit for remembering anecdotes and details. "This is such an outrageous tale," we think, "there's no way I could ever forget it." Oh, yes you can.

In the late 1980s, one of us was smart enough to sit her grandmother down for an audiotaped interview. Nana was 90 years old at the time, but in surprisingly good health. The focus of the interview was to get her to recount all the standard family stories she had told through the years—the time she literally ran into J. P. Morgan in New York City (and he had been charmed because she was such a pretty young thing), the relative who went across the Oregon Trail at the age of 18 and pregnant, the grandfather with wanderlust who used to sail back to Ireland without warning whenever the mood struck him, and so forth. Sadly, Nana passed away a mere 3 months later. And while the "if only I had . . ." self-torment had been narrowly avoided, it was amazing how much of the content of the tape had vanished from memory before listening to it again a few years later—in spite of the fact that these were the tales we had been raised on, the ones that were almost annoying in their repetition.

So even if you have a remarkable memory and are noted for your instant recall of names, dates, and figures, please record your findings. Fortunately, this is easy to do with just a few simple tools. All that's necessary is to familiarize yourself with a handful of basic genealogical forms that will help you systematically organize all the information you're uncovering. Better yet, if you're computer literate, invest a small sum in software (most are less than $100), or use an online family tree–building tool (some are free) to help you capture all the details. Using such software, you can enter the information once and automatically generate all the different reports you need. Most of these packages also include notes features, which allow you to record all those stories you've been told. Appendix A provides resources for forms and software.

Chart Your Course

While there are a variety of standard genealogical forms, two types of charts will be especially helpful in your quest to understand and learn about your genetic roots. (See Appendix A to learn more.) These are the **pedigree or ancestral chart** (a form showing the direct-line ancestors of a particular individual) and the **descendancy chart or descendant tree** (a chart in which a selected ancestor appears at the top, and all his descendants are situated in successive generations in rows below him).

THE POWER OF PEDIGREES

Figure 1-1 shows a portion of the pedigree or ancestral chart for Petrus Smolenyak. Starting from the first box on the left, we find

Figure 1-1: Example of a pedigree or ancestral chart

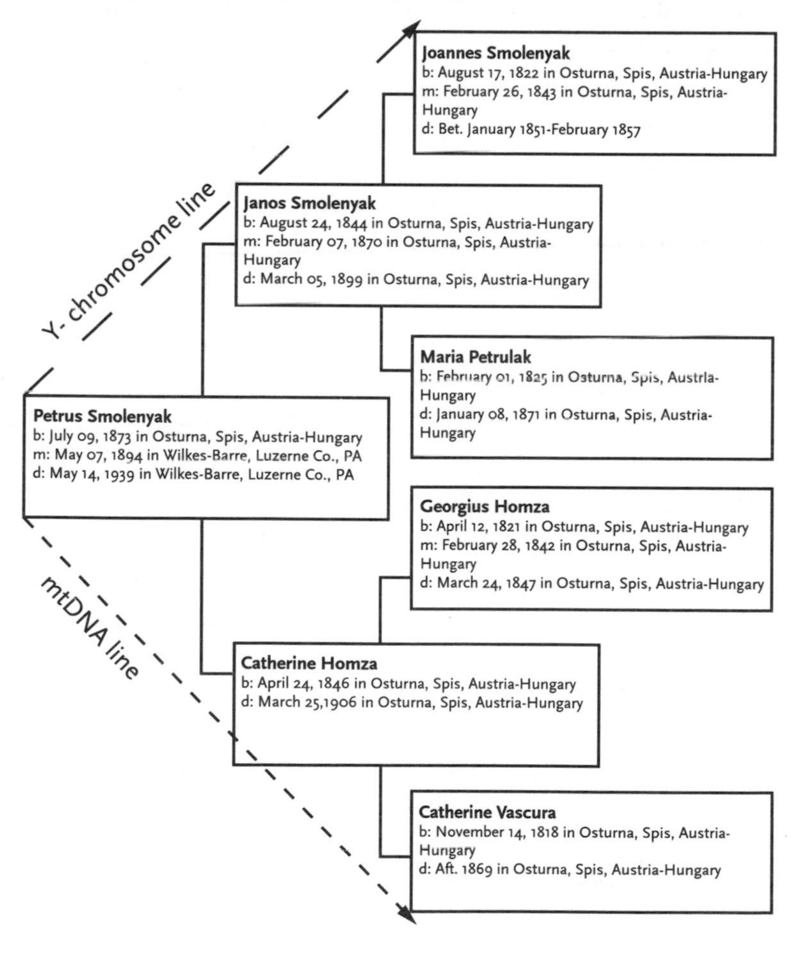

facts about his birth, death, and marriage. Shifting our attention to the middle section, we can discover details about his parents. Moving right once more, we find boxes with information about his grandparents. This is a typical ancestral chart, although we've abbreviated the one here to three generations to make it easy to view. We should also point out that pedigrees are sometimes presented in a vertical arrangement. (You can skip ahead to Figure 2-1 in Chapter 2 for a moment if you'd like to see an example.)

Pedigrees are helpful in genetic genealogy because they make it easier to understand and explain how particular DNA tests function in the world of family history. Some people mistakenly believe that taking a DNA test will magically reveal everything they've ever wanted to know about their roots, but the most popular tests, Y-chromosome and mitochondrial DNA (mtDNA), only answer questions about a portion of your family tree.

While we'll be exploring all of this in great detail in the coming chapters, we'd like to quickly point out that Y-chromosome testing essentially maps the top line of your pedigree (as can be seen in the dotted arrow heading toward the upper right in Figure 1-1). In a genealogical sense, this is very convenient because it corresponds to the surname associated with this uppermost line. This also means, however, that taking a Y-chromosome test will only answer questions pertaining to this particular surname or paternal branch of your family tree.

Similarly, mtDNA testing maps the bottom line of the same chart (as can be seen in the dotted line pointed toward the lower right in the same illustration). Unfortunately, the surname represented is not consistent since women in most cultures have traditionally taken their husband's name upon marriage. In this illustration, for instance, we see a Smolenyak child with a Homza mother and a Vascura maternal grandmother—three surnames in three generations. So mtDNA testing does not provide the same built-in convenience of a DNA-surname match that Y-DNA testing does, but it can shed light on the maternal branch of your family tree.

Depend on Descendancy Charts

The other type of chart that's especially useful for genetealogy is the descendancy chart or descendant tree, as seen in Figure 1-2. In this case, we are able to see descendants of the focus individual (often referred to as the *ancestor of interest*), Joannes Smolenyak, displayed generation by generation underneath him. While such charts are a staple in all genealogical research, they are used with great frequency for DNA purposes because of their value in illustrating the path of a given DNA line.

It is not unusual, as in this tree, to see males represented as squares and females as circles. This convention, borrowed from the realm of genetics, makes it easier to follow a desired DNA line in either direction—down from the ancestor of interest or up from someone living today. And while this sample chart has been simplified, typical descendant trees can get quite busy and large, so another habit many genetealogists have adopted to make the DNA trail even more apparent is to shade or otherwise highlight the entries of particular interest. For instance, all the men in the shaded boxes in Figure 1-2 would be expected to have the same Y-DNA.

Figure 1-2: Example of a descendancy chart or descendant tree

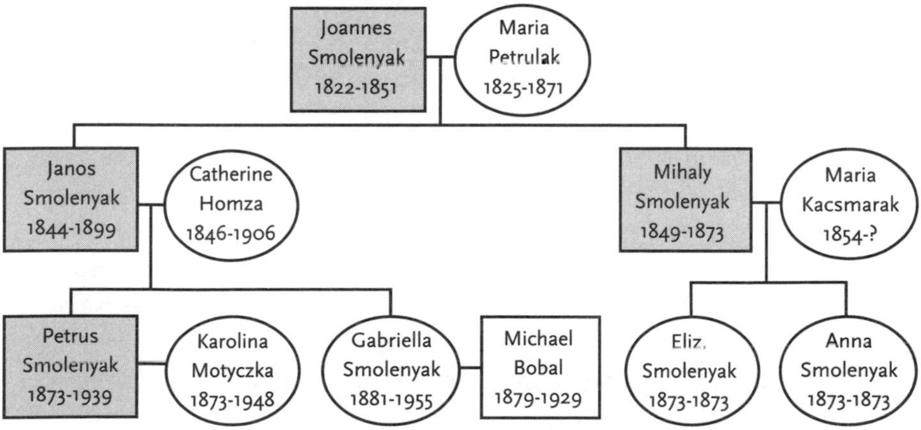

A variation on the descendancy chart is the *outline descendant tree* (see Figure 1-3). This presents the same information, but in a collapsed, text-only format. Starting with a selected ancestor at the top of a page, all descendants are listed beneath. Each successive generation is indented slightly more from the left margin than the one that precedes it. Every individual has his own line (which may wrap depending on its length) and is followed by listings for spouses and children. While many find standard descendancy charts easier to grasp, they have a tendency to become rather wide after a few generations or when dealing with large families. This can result in awkward, oversized printouts or a lot of left-right scrolling for the computer user. Outline descendant trees may take a little more effort to absorb, but they convey the same data using a style that fits on standard-size paper or a computer screen, so they are also frequently used for genetic genealogy.

Don't Skip Generations

In Beth Maltbie Uyehara's amusing essay "The Ten Commandments of Genealogy" (from her book *The Zen of Genealogy*), she insists, "Thou shalt never leap back a couple of generations just because it sound-eth like fun." When it comes to genetealogy, we're going to ask you to fervently obey one portion of this commandment, but disregard another aspect.

In traditional genealogy, we're trained to start with ourselves and work back through the generations, which is why Beth has specified that you should not leap *back*. But conducting a DNA project often requires the reverse—starting from a point in the past and working forward in time. In order to solve your particular mystery, for instance, you may find yourself trying to locate descendants of an English immigrant who came to the United States around 1800. In this case, you would need to start with the immigrant, identify his children, and progress toward the present day through his grandchildren, great-grandchildren, and each subsequent generation in the hope of finding someone alive today

Figure 1-3: Example of an outline descendant tree

DESCENDANTS OF JOANNES SMOLENYAK

1 Joannes Smolenyak b: August 17, 1822 in Osturna, Spis, Austria-Hungary d: Bet. January 1851–February 1857

.. [+]Maria Petrulak b: February 01, 1825 in Osturna, Spis, Austria-Hungary m: February 26, 1843 d: January 08, 1871

.........2 Janos Smolenyak b: August 24, 1844 in Osturna, Spis, Austria-Hungary d: March 05, 1899

...............[+]Catherine Homza b: April 24, 1846 in Osturna, Spis, Austria-Hungary m: February 07, 1870 d: March 25, 1906

....................3 Basilius Smolenyak b: February 01, 1872 in Osturna, Spis, Austria-Hungary d: June 02, 1872

....................3 Petrus Smolenyak b: July 09, 1873 in Osturna, Spis, Austria-Hungary d: May 14, 1939

..........................[+]Karolina Motyczka b: October 19, 1873 in Barwinek, Austria-Hungary m: May 07, 1894 d: July 30, 1948

..................................4 Peter Smolenyak b: March 09, 1895 in Wilkes-Barre, Luzerne Co., Pennsylvania d: January 27, 1947

......................................[+]Bertha Piska b: 1900 in Wilkes-Barre, Luzerne Co., Pennsylvania d: February 25, 1954

with the appropriate DNA for your research. Because this is backward from the approach used in most traditional genealogy, we've dubbed it *reverse genealogy,* and you'll hear us use this phrase repeatedly in this book, especially in Chapter 8, Finding Prospects.

But just as with traditional genealogy, it is critical to never skip generations. With so many opportunities in our lives for instant gratification, we've become accustomed to fast results and have come to expect our roots served up the same way. So it's only natural that many of us embarking on a genealogical quest will be tempted to leapfrog across generations in the course of our research.

In reverse genealogy, this will most frequently manifest itself in the urge to borrow several generations of information found in a published genealogy or online family tree. Why trouble to methodically work your way through each generation when you can sail through three or four by simply plugging in the data you found in some great book or on the Internet? Remember what we said a short while ago about not believing everything you hear or read? That applies here. Unfortunately, there's a lot of wishful thinking, poor research, and even intentional deception built into existing genealogies.

Back before genealogy became a hobby for all of us, many of the first to dabble in it a century or more ago did so essentially for bragging rights—to prove illustrious roots. And if that meant a little stretching of the truth, a little force-fitting to establish a royal connection, what was the harm? Little did they imagine at the time that genealogy would blossom in the way it has and that hundreds or even thousands of descendants would read this work and accept it as truth—in spite of red flags such as men having children when in their 80s or girls becoming mothers at the age of 10. And while most of us today are after the truth and are just as delighted with our rebel and rogue ancestors as with any aristocrats in the family tree, the Internet has made it possible for us to spread misinformation—accidentally or otherwise—at lightning speed. So if you accept what you discover online or in published sources without question and scrutiny, you are almost certain to inadvertently inject some misinformation into your family tree. In fact, genetic genealogy can be a powerful tool for exposing the fiction behind some flawed research that has been widely spread over the years. But assuming that your primary interest is to learn the reality of your family's past—whether you're doing traditional genealogy or working in reverse for a DNA project—it's best to make it a habit to advance one generation at a time, confirming each parent-to-child link.

It's Not All on the Internet

Yes, the Internet is a vast resource that has revolutionized the world of genealogy, but some folks are under the mistaken impression that everyone's family history has already been researched, and all they need to do is type in their name, and their roots will instantly appear. While it's true that more and more of us can quickly find bits and pieces of our family's past sprinkled across the Internet, those who expect to find it all will be disappointed.

And even if you're fortunate enough to find information online, you should only use it as a starting point because you never know how accurate it is until you follow the paper trail yourself. We could create a Web site for a fictional family of your surname and upload it to the Internet in a matter of minutes. Not only would no one stop us, but in perhaps 6 months or so, there's a very good chance that others researching the same name will have "borrowed" information from our site and inserted it into their own genealogies, which they will then post online where still others can "borrow" it.

This is why misinformation in a cyberspace environment is so contagious and why it's so important to do your own research—and that means more than surfing the Internet. If you're prepared to invest in a few subscriptions, you can access many records online, such as most U.S. federal census records and a growing selection of passenger arrival records. (If you find the cost too high, you can frequently access the same resources for free through your local library.) But many records are created locally and will only be of interest to those with roots in that geographic vicinity. Consequently, it's necessary to consult local resources to find them. If you're lucky, you may be able to find what you're seeking online, but more than likely, you will only find pointers to the repositories that hold the documents you need.

It's always worth searching the catalog at www.familysearch.org, for instance, to see if the Family History Library in Utah has any records for your place of interest. If so, you can hire a researcher in Salt Lake City, plan a research trip there, or order selected microfilms to be delivered to the Family History Center closest to you for a 3-week loan period. The more than 3,400 centers located around the world are all listed on the same site, so you can search it to find the one closest to you.

You'll also want to bookmark www.usgenweb.com and www.rootsweb.com, where you can frequently find excellent state and county-based sites that may house treasures such as searchable archives of marriage indexes or cemeteries for the county your family lived in for a century or more. The content of these sites, however, is extremely uneven. Some areas have deep resources to wade through, while others have a bare minimum. For the most part, they will at least provide contact information for public libraries, local newspapers, county courthouses, and relevant genealogical and historical societies. You may be able to follow a link to them and send an e-mail, but you may—shudder!—have to write a letter or take a field trip to research onsite.

Please don't let this deter you. It's just part of the genealogical quest. After all, it wouldn't be much fun if you really could just type your name in and find everything laid out for you. Discovering the clues that have been scattered around the county, state, country, or even the globe makes it much more gratifying when you find one of the missing pieces of your family history puzzle. And for reasons we can't quite explain, there's something a little more exciting about getting that long-sought document in the mail or plucking it from the obscurity of a courthouse basement than stumbling across it online. In fact, we suspect it won't be long before you find yourself addicted to the thrill of the hunt!

Surround and Conquer

When we're researching a particular ancestor, our tendency is naturally to fixate on the paper trail generated by that individual. But it seems to be an especially common aspect of Murphy's Law that the ancestor you're seeking will be the one whose life was completely undocumented or whose records were destroyed in floods and courthouse fires. Or maybe the ancestor was a female in times when women left relatively few traces.

For this reason, it's good genealogical practice to research not only your direct line forebears but also *collateral lines* (that is, descended from a common ancestor, but through different lines, such as cousins) and even others connected to them by marriage or casual association. In short, find the answers you seek about your ancestor by looking into the lives of those who surrounded her. If you can't find out what Maria Dyer's married name was, maybe her brother's obituary will furnish the missing detail when it lists his survivors. If you can't find out where Bridget Kelly came from in Ireland, perhaps a little research into those who are buried in the same cemetery plot with her will turn up some clues. Almost all our ancestors' paper trails are spotty, but paying attention to siblings, cousins, and even neighbors and other associates (fellow members of church, community, and social organizations) can often fill your gaps.

This surround-and-conquer approach can be especially handy for genetic genealogy because lines of people with the appropriate DNA for your research may die out over time (more on this in Chapter 8, Finding Prospects), but if you're willing to pursue collateral lines, there's a very good chance that the DNA you need is still available through some distant cousins.

Now that you've soaked up some genealogical fundamentals, let's move on to the other component of genetealogy—genetics.

2

Genetic Essentials

Are you a geneticist? Don't be too quick to answer no. Almost certainly you have practiced the trade from time to time. Using only your eyes as tools, you've made plenty of genetic observations. Have you ever taken sides in a friendly argument over whether a newborn baby looks more like his father or his mother or even his great-uncle John? Have you perhaps envied a prominent sports figure who seemed to get a head start by picking the right parents? Have you counted how many of your uncles lost their hair at an early age or noticed that some children in a family have blue eyes while others have brown eyes? Then you are thinking like a geneticist.

In this chapter, we'll survey the field of genetics and pack an introductory course's worth of information into a few pages. It's a lot to learn, but you don't need to master it all (we promise—no pop quizzes!). In fact, you'll still be able to benefit from DNA testing even if you can't rattle off the meaning of all the terms. And you can always refer to the glossary or back to this chapter later if you feel the need.

Classical Genetics

Classifying and counting are the methods used in "classical genetics," sometimes called Mendelian genetics. The field of classical genetics began with the Austrian monk, Gregor Mendel, who lived from 1822 to 1884. Year after year, crop after crop, Mendel classified and counted (and counted and counted) the peas in his monastery garden—tall peas and short peas, plants with red flowers and plants with white flowers, pods with wrinkled peas and pods with smooth peas.

As Mendel experimented with many combinations of characteristics, he began to see patterns in the way traits were passed on, sometimes disappearing in one generation and reappearing in the next. He noticed that one alternative was always dominant: If he crossed pure-breeding round peas with wrinkled peas, the offspring were always round.

But if he crossed those round offspring with each other, a few of the next generation would be wrinkled again. It was as if the instructions for creating wrinkled peas were lying dormant, waiting for their chance to see the light. At the same time, if the two parents were wrinkled peas, they never produced offspring with round peas.

Mendel was the first to use the words *recessive* and *dominant* to describe this effect. The recessive version could only appear when the dominant trait was absent. This implied the existence of two versions of each characteristic, one inherited from each of the parents. These alternative versions are called *alleles.* (For a memory jogger, remember the first two letters in the words alternative and allele.)

Although a dominant trait masks the presence of a recessive trait, it is not necessarily the superior version, nor is it the one that is more likely to be passed on to the next generation. The offspring have a 50-50 chance of inheriting one allele or the other from the parent, so it's as random and unpredictable as a coin toss. That's what makes siblings so different—any child may or may not inherit a particular allele from his parents.

Mendel also concluded that each characteristic was independent: He could eventually create any combination of height, flower color, and pea shape by picking the right parents. This became known as the law of *independent assortment*. Furthermore, it did not matter whether the traits came from the male parent or the female parent.

Today we know that some characteristics are linked together (when the genes lie close together on the same chromosome, they are inherited together as a package), that some characteristics have more than two alleles (there are three alternatives for blood type, called A and B and O), that some characteristics are codominant (one person can have blood type AB because the two alleles are both expressed), and that the sex of the parent sometimes influences whether certain traits are expressed *(sex-linked)*. But by sheer luck, Mendel had avoided some of these more complicated modes of inheritance.

Types Galore—Blood Types, Phenotypes, and Genotypes

The methods of classical genetics are still in use to this very day, but they are limited to traits that can be observed or measured in some way. Geneticists call this the *phenotype* (from the same word root as phenomenon, something which is apparent to the senses). But outward appearances are just the tip of the iceberg. Are there similarities and differences that can be identified with more refined tools? The answer is yes; blood type is one example.

You may know your ABO red blood cell phenotype if you've ever donated blood or had a blood transfusion. Your phenotype determines how your body will react when it encounters blood cells from another person. The gene for blood types A and B creates two different kinds of sugar molecules, which attach to the surface of the red blood cell. If your body creates the A version, the B version will seem like a foreign object, which the immune system should destroy. People with type O cannot create either sugar molecule, so both A and B type cells will be treated as invaders in their bodies.

We now know that even people who have the same phenotype—for example, blood group A—may differ at a deeper level. One person with blood type A may have inherited the A allele from both parents (A-A), but another person may have inherited the A allele from one parent and the O allele from the other parent (A-O). Since A is dominant over O, the A-O blood reacts as A-A blood in transfusions. The phenotype is A for both people, but the two people have a different *genotype,* that is, a difference that can only be detected by molecular methods.

Molecular Genetics

Professional geneticists have new tools at their disposal that can delve right down to the molecular level, to the code that determines the traits. Many of these methods were developed during the course of the Human Genome Project. It's actually possible to take a sample from any kind of cell in the body, even a cell rubbed off the inside of your cheek, and determine the red blood cell type by looking at the DNA code. The power to reach the molecular level—to know at the most fundamental level how people are similar or different—is what makes genetealogy a viable proposition.

Nowadays, these tools are not limited to research laboratories. We can all order tests from commercial outfits that cater to the genealogical market (which we mostly refer to simply as "testing companies"). The results of these tests are much more abstract than descriptions of traits. In fact, they have nothing to do with traits at all. You may see a string of numbers or letters, such as 14–12–24–11–13–13 or AGCTTACTTAG. That may seem about as interesting as a bar code or a novel typed by a monkey on a keyboard with four letters. But these seemingly dry-as-dust results record the history of your ancestors and come alive as you learn how they can connect you to your close and distant kin.

Much of the scientific vocabulary has crept into widespread use, such as DNA, chromosomes, and genes. Most of us have an idea of what these words mean, but let's take a little time to explore each of them—and while we're at it, get down to some nitty-gritty details that will help you understand how DNA testing works for genealogy.

DNA Is the "Molecule of Life"

DNA is short for DeoxyriboNucleic Acid (dee-ox-ee-rye-boh-new-clee-ic acid). It's one of the most valuable components of your body, even in a purely monetary sense. If you could extract all of your DNA, it would weigh 7.5 grams (about a quarter of an ounce), and you could get more than 9 million dollars for it, as Patrick Di Justo reported in *Wired* magazine. The molecule is made of cheap ingredients: the atoms carbon, hydrogen, oxygen, nitrogen, and phosphorus. It's the way those ingredients are assembled that gives DNA its special status.

But even if you could extract all that DNA without damaging your body, you wouldn't live long. DNA's job wasn't finished when you matured into an adult. It is busy every second of every day, instructing blood cells to make hemoglobin to deliver oxygen to your cells, telling your pancreatic cells to make more digestive enzymes when food leaves your stomach, and remodeling your bone cells to keep them sturdy.

The Basics of Bases

DNA is a very long, very skinny molecule that strings together simple units called *bases* (or sometimes *nucleotides*). If you could stretch it out, the DNA from just one cell would be about 6 feet long, but it is coiled and twisted until it can be packed inside the cell. The bases—called adenine, guanine, thymine, and cytosine—are abbreviated A, G, T, and C. Remember the science fiction movie *Gattaca*? The title was constructed from these four letters.

The simplicity of the DNA molecule led early scientists to dismiss it as the source of all of life's complexity. But Morse code can construct any word in the dictionary or even works of literature by using just two signals: dots and dashes. A very detailed owner's manual can be written with four bases—and fortunately for us, a cell knows how to read and follow these instructions step-by-step.

The goal of the Human Genome Project was to establish the *sequence* (order) of all of the DNA bases, some 3 billion of them. The results are written in an abstract shorthand; for example, AGCTTTACGGA. By the year 2003, the effort was essentially complete. (Anyone who wants to read the "magnum opus" written with the letters A, G, T, and C can download chapters—one for each chromosome—from publicly accessible databases, such as www.ncbi.nlm.nih.gov/mapview/map_search.cgi?taxid=9606.)

The DNA Copy Machine

The DNA molecule chains millions of bases into two long strands that spiral around each other in the famous double helix shape. The bases have three-dimensional structures that nestle together like pieces in a jigsaw puzzle—if the right pieces are adjacent to each other. Adenine fits with thymine, and cytosine fits with guanine in *complementary pairs.* (If you want to remember these pairings, the letters C and G both have curves, while the letters T and A have all straight lines.)

These C-G and T-A pairs are not tightly bound and can be eased apart when the cell makes new copies of the DNA molecules, just before it divides in two. Each separated strand finds complementary bases and lines them up with the help of an enzyme called *DNA Polymerase.* Each copy acts as a template for the next copy, like looking at a mirror image of a mirror image. Scientists have learned to mimic the action of the cell in a process called *Polymerase Chain Reaction,* or *PCR* for short. PCR doubles and redoubles the DNA in a sample many times, which helps us read DNA from very small samples such as those used for genetealogy projects.

Chromosomes Are Packages of DNA

The DNA molecules are packaged, along with some proteins, into individual structures called *chromosomes.* The chromosomes

Chromosomes under the Microscope

The chromosomes are so jumbled together that it was hard to even count them in the early days. For many years, biology textbooks gave the number of chromosomes as 48. In 1956, Tjio and Levan published photographs (called karyotypes) that clearly showed 46 chromosomes. The first scientists to count the human chromosomes had made a mistake, and other researchers who could find only 46 chromosomes in their slides assumed they just couldn't spot the other two in the clutter!

are found in the *nucleus* of the cell, a central region separated by a membrane from the remainder of the cell contents.

The word *genome* refers to the whole set of chromosomes. All chromosomes—*except* X and Y—come in pairs (which is why you'll sometimes read about 46 chromosomes and sometimes about 23, depending on the context). These pairs are numbered 1 to 22 and are sometimes called the *autosomes.* You receive one copy from your father and one copy from your mother.

The X and Y chromosomes are called the *sex chromosomes* because they determine gender. Two X chromosomes produce a female, while an X and Y combination produces a male. It is the father who determines the sex of a child. The mother, with two Xs, can only provide an X, so if the father passes along his Y chromosome, the infant will be a male. But if he passes along his X, the child will be a female.

If You're a Man

Let's look at this from your perspective. If you're a male, you received your X chromosome from your mother and your Y chromosome from your father. Where did your father get his Y chromosome? There's only one possibility—*his* father. Your Y chromosome came from your father's father's father's father's father . . . a straight paternal line stretching back into the mists of time.

Figure 2-1 illustrates the path of the Y chromosome on a pedigree chart arranged vertically. The starting person (you, if you're a man) is shown at the bottom, then the two parents (the father in the square on the left and the mother in the circle on the right), the paternal and maternal grandparents, and the eight great-grandparents at the top. The Y chromosome is always on the left-hand side in this style of chart.

Figure 2-1: Inheritance of Y chromosome

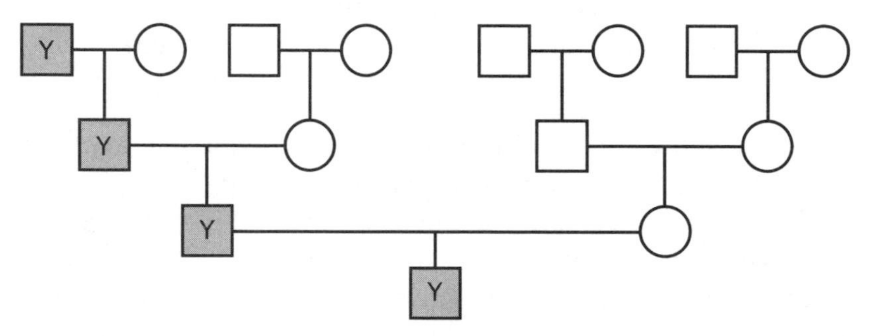

When you study the Y chromosome, you know exactly where it came from. You may not know the name or the date or the place, but you know the slots on your pedigree chart where it belongs, and it conveniently follows the surname line in many cultures! The next chapter will demonstrate how surname projects can use the Y chromosome.

But where did your X chromosome come from? Why, from your mother, of course—but where did *she* get it? She has two Xs, and she may give you the one she received from her father *or* the one she received from her mother (or more likely, a patchwork of the two chromosomes, through a process called *recombination*). Thus, as you can see in Figure 2-2, your X chromosome can have DNA from one or two of your four grandparents: your mother's father and your mother's mother. Going back another generation, it can have DNA from any three (but no more than three) of your eight great-grandparents. If there is an X-linked condition in your family, such as colorblindness, it must have come from one of the shaded circles or squares in Figure 2-2.

Figure 2-2: Inheritance of the X chromosome in a male

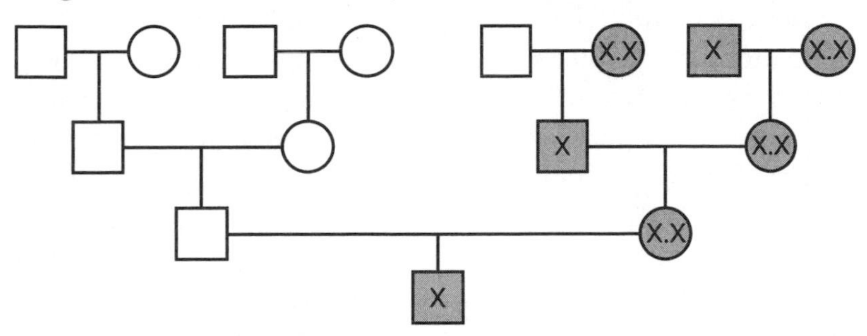

If You're a Woman

If you are a female, you have two X chromosomes, so you can have representation from more of your ancestors: three of four grandparents, five of eight great-grandparents, and so on. The total number of ancestors doubles with each generation (2, 4, 8, 16, 32, 64, 128), but the number of ancestors who could contribute an X increases more slowly, as seen in Figure 2-3.

Figure 2-3: Inheritance of the X chromosomes in a female

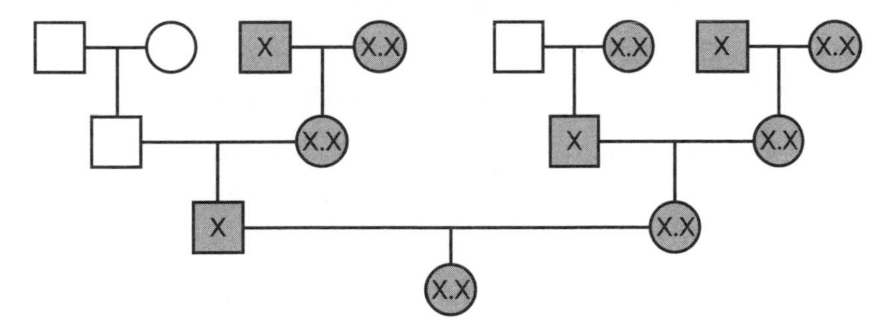

We slipped a tiny weasel word into the preceding paragraphs, which might have escaped your notice. Consider the implications of the word *can* in the phrases: "you *can* have representation from . . ." Those numbers are the maximum possible. By the time you reach the level of your great-grandparents, much of their DNA has dropped out of the picture. Only 50 percent of a parent's DNA can be passed on to the next generation. If you are looking for a particular snippet of DNA from your mother's mother's father's mother's X chromosome, you may very well be out of luck. Every snippet does come from some ancestor—but because of shuffling and recombination, you can't tell which one. This same principle applies to autosomal (not on the X or Y chromosome) DNA as well, although we have concentrated on the X to emphasize that it does not behave the same way as Y, as many assume. You may sometimes encounter a phrase that underscores the special properties of the Y chromosome: non-recombining Y (NRY).

Women are often disappointed to learn that their DNA cannot be used in surname projects, because they have no Y chromosome. On the surface, this effectively cuts women out, but women can participate by proxy. If they're interested in learning about their maiden name, for example, they can talk a brother, cousin, father, or uncle into testing in their place. And there's a consolation prize—another type of DNA passes down through the egg alone and consequently follows the female line. It's called mitochondrial DNA, and we will dedicate an entire chapter (Chapter 4, Maternal Legacy) to this tiniest of DNA molecules and how it can be used to shed light on your roots.

Genes Are Stretches of DNA with a Mission

Curiously, the word *gene* is getting harder and harder to define, but a clear definition was essential when scientists from the Human Genome Project decided to wager on the total number of genes discovered by the year 2003, when the complete DNA sequence would be reported. Wilhelm Johannsen coined the term in 1909 as a short, simple way of referring to the fundamental unit of heredity. The word is derived from the Greek *genos,* which encompasses the idea of birth and extends to related concepts such as family groups. You can spot the root in many words: progenitor and progeny, indigenous, genesis, generate, and generation—and of course, genetics, genealogy, and now genetealogy!

The basic concept remains the same: The gene is the unit of heredity. The biochemical function remains the same: The order of bases in a gene prescribes the order of amino acids in a protein. That's why DNA is often referred to as a recipe, blueprint, or parts list. Just about everything in your body involves proteins at some level. There are structural proteins such as collagen, enzyme proteins that expedite all the chemical reactions in your body, regulatory proteins that turn hormones on and off, antibody proteins for immune functions, and countless other proteins for thousands of familiar and cryptic functions. All these different proteins are just different arrangements of the twenty-some different amino acids, which twist and turn and bend into different shapes to do their jobs.

There are more than 1,000 genes on the X chromosome, while the count for the Y chromosome in the year 2003 stands at just a fraction of that: 27. The genes on the X chromosome have little or nothing to do with sexual characteristics. They cover a broad range of structure and function, much like any of the autosomes.

The Y chromosome acts like a switch—if it is present, the baby will be a male. Genes restricted to the Y chromosome could hardly be essential for life and health, else the female of the species would disappear. Classical genetics has never identified any traits or diseases linked to the Y chromosome, so there is no need to fear that sharing DNA results will impact the ability to obtain health insurance.

Oh—were you wondering about the bet about the total number of genes? The range of bets was 27,642 to 152,478. The final answer was somewhat arbitrarily set to a mere 21,000, but there's plenty of scope for more argument!

One Man's Junk Is Another Man's Treasure

Did you know that there are long stretches of DNA between genes with no known function? In fact, an astonishing 95 percent or more of the human genome does not code for anything. The term *junk* DNA is used somewhat tongue-in-cheek. A more neutral phrase is *non-coding DNA*, because it may very well have purposes that we don't understand yet.

Since junk DNA does not code for anything, it has no effect on personal traits or medical conditions. Yet it has great utility for another purpose. Junk DNA is recording your ancestral history just as clearly as your genes—perhaps even more clearly since it has no selective effects on survival. Mutations accumulate freely, preserving the evidence of events that happened thousands of years ago. Population geneticists have used junk DNA to study the migrations of ancient peoples "Around the World," the title of Chapter 5.

All of the currently available tests for the genealogical community use *markers* from non-coding DNA. That is reassuring for people who worry that participating in DNA projects might affect their ability to obtain health insurance. So what can we learn from this meaningless junk DNA? The answer is simple but profound: We are all related, but closely related people will have more similarities than distantly related people. Molecular genetics has indeed opened a treasure chest for our inspection!

Marking Up the Map

The phrase "genetic marker" includes an ordinary English word. Gravestones, mileposts on highways, seed packets at the ends of garden rows, and the X on a treasure map are all markers that help us pinpoint the presence and location of something. Geneticists use the Latin word *locus* for a position on a DNA molecule. (The plural is loci, pronounced low-sigh.)

A genetic marker is a distinctive feature on the DNA molecule that allows us to flag a particular locus and study it in more detail. We have to know something about the DNA sequence in the vicinity to start the procedure though. Thanks to the investment in the Human Genome Project, we as consumers have inexpensive access to numerous genetic markers. We will discuss different types in more detail later, in the chapters that describe the various tests that are available. In fact, genetic markers are most interesting when they alert us to these individual differences. Genetic differences are sometimes called *polymorphisms* (from the word roots poly = many and morph = form). If there are polymorphisms, something had to change, which brings us to the subject of mutations.

Mutations

If your only exposure to the word mutation is from science fiction or horror films, you'd be entitled to worry about participating in DNA projects that proclaim your mutations to the world. In fact, the word simply means a change, and changes in DNA (like changes in life) can be good, bad, or indifferent. Since the genealogical markers use junk DNA, changes are automatically indifferent.

What causes these changes? There are a few agents (called *mutagenic,* giving birth to mutations), such as high-energy radiation, free radicals, viruses, and certain uncommon chemicals such as mustard gas. These *induced* mutations are relatively rare, however. Most mutations are *spontaneous*—an "oopsie" caused when DNA polymerase doesn't do a perfect job. It is amazingly accurate; it makes a copying error perhaps once in every fifty million bases. Still, we should be grateful that polymerase isn't perfect—without some mutations, life would never have evolved! And without mutations, we'd have no basis for genetealogy because it's these very rare changes that we use for purposes of comparison.

Mutations can thus be categorized as spontaneous or induced, based on causative factors. Another way to categorize them is based on where the mutations occur, whether they are *somatic* or *germ line.* Somatic mutations occur somewhere in the body, and since they can't be passed on to the next generation, they're not much use for genealogy.

Germ line mutations occur when the egg and sperm are formed. (The word germ is meant to convey a small structure that turns into something big, as in "germ of an idea".) Germ line mutations are heritable, making them the kind that can be used to track connections between generations and, therefore, for genetealogy.

People sometimes ask if somatic mutations will affect their DNA tests. For example, radiation to the head and neck region might cause mutations in the cells inside the cheeks, which are used to submit DNA samples to testing laboratories. It is a theoretical possibility, but not a practical concern. Thousands of cells are combined in the test tube, and a mutation that hits one spot on one DNA molecule is unlikely to affect all the cells the same way.

Enough of the science lesson. We suspect that what you really want to know is how all this pertains to your roots. . .

Part 2

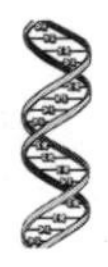

Testing Options Explained

3

Male Bonding: Y Chromosome

If you are a man, you have in your possession a Y chromosome that whispers a coded message from your distant ancestors. Don't skip this chapter if you're a woman, though! You have as many forefathers as any man, and even men must call upon collateral relatives to learn about some of their ancestral lines.

The Source of All Y

If we had a time machine, we could trace the Y chromosome of every man living today back to one man. This is mind-boggling to contemplate, but if we take the journey step-by-step, there is no other destination. Consider the set of all males living today who have no sons of their own. That's a very large number, perhaps approaching a billion. They all inherited their Y chromosome from their fathers. How many fathers are there? If every man were an only son, the number of fathers would be identical.

But some men have brothers, so the number of fathers is bound to be smaller.

The same reasoning applies to the fathers of those fathers. The total number shrinks a little bit more with each generation back. In some families, there is only one grandson carrying on the family name, while in other families, several first cousins may converge on one grandfather. (See Figure 3.1 for a diagram showing this process of *coalescence.*) The exact count may be hard to estimate, but the number can never grow, for no man has two biological fathers.

Y-Adam

Generation by generation, the tally dwindles, from millions to thousands to hundreds. Eventually all the branches lead to one man, the *Most Recent Common Ancestor (MRCA)* of all men in the straight paternal line. He was a real person, not an abstraction. He is sometimes nicknamed Y-Adam, but that moniker can

Figure 3-1: Coalescence to the Most Recent Common Ancestor (MRCA)

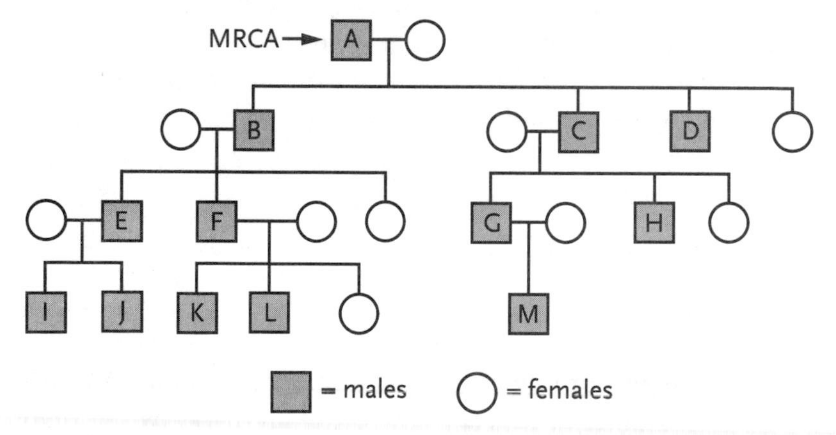

E is the MRCA of the brothers I and J, while F is the MRCA of the brothers K and L. The four cousins (I, J, K, and L) coalesce to their MRCA B. C is the MRCA of M and his uncle H, making A the MRCA of all the males on this diagram.

cause some confusion if it is interpreted too literally. He was not the first man, for Y-Adam also inherited his Y chromosome from his father. He is simply the most *recent* Y-line ancestor.

If Y-Adam was not the first man, neither was he the only man living at the time. Y-Adam had many contemporaries, some of whom might have even more living descendants altogether. But some of those fellows had only daughters, and their Y chromosomes died out immediately. Others had sons who kept their Y chromosomes perking along for a few more generations, or occasionally even for thousands of years, but eventually there were no sons to carry on the Y in all those other branches.

What do we know about Y-Adam? He was almost certainly born in Africa. The date is not nearly as clear, but several different methods calculate that he was born less than 100,000 years ago, perhaps as recently as 60,000 years ago. By definition, Y-Adam had at least two sons, who also had one or more sons, who also had one or more sons. Some of his descendants eventually left Africa and migrated to all the corners of the world. Spencer Wells's vivid book *The Journey of Man: A Genetic Odyssey* is a veritable travelogue that follows their paths.

Practically Perfect

If every father down through the ages had duplicated a perfect copy of his Y for his sons, every man in the world today would have an absolutely identical chromosome. In fact, as recently as 1995, a search for polymorphisms in a worldwide sample of 38 men came up empty. Only a small portion of the chromosome was examined, but it appeared that the Y would be useless for genealogy.

Since then, scientists working on the Human Genome Project have sequenced more of the Y chromosome and developed new techniques to pinpoint slight differences. Would the pendulum swing the other way? If every father tweaks his Y just a little bit before passing it on to his sons, even brothers will be somewhat different upon close inspection. First cousins would be more variable still, and it would become difficult to detect similarities in more distant relatives. Genealogists are like Goldilocks—we want to see changes that are not too slow and not too fast, but just right. The optimal *mutation rate* would preserve the same signature for maybe 10 to 20 generations, back to colonial times in America or the origins of surnames in Europe. The Y chromosome was not designed with genealogists in mind. Are we asking for the impossible?

Amazingly, there is a category of genetic mutations that tweaks the Y chromosome in just this way. They're called *Short Tandem Repeats (STRs)*. Certain junk DNA segments on the Y chromosome seem to stutter: A *short* pattern (often two to five bases in length) is *repeated* a number of times in a row (in *tandem*). For example, the four-base pattern GATA might be repeated five times: GATAGATAGATAGATAGATA. Very occasionally, the enzyme responsible for duplicating the DNA on a chromosome loses its place, and the copy ends up a little shorter or longer, with four or six repeats instead of the original five. Scientists measured some STRs in many father/son pairs and concluded that this mutation, a simple little copying glitch, occurs two times out of a thousand on the average. STRs are very polymorphic. They typically come in five to 10 alternative versions (alleles), based on the number of repeats. For example, a typical range for a marker called DYS19 is 12 to 16 repeats, and the range for another marker DYS390 is 21 to 26 repeats.

What's the Catch?

Mutations don't occur like clockwork. They are random events—we cannot predict when they will happen. The original value may be preserved for hundreds of generations. On the other hand, two brothers may end up with different values, for a mutation has to happen sometime. Clearly, we can't say exactly when a common ancestor lived just by comparing the results of a test on one marker. There is a broad range of possibilities.

The picture will be somewhat clearer if we use a suite of independent STR markers, perhaps 10, or 20, or 30, or so. If two men are brothers, we would expect them to match on most of the markers, if not all. If two lines diverged from a common ancestor a thousand years ago, we would not be surprised to see a mismatch show up on one or another of the markers.

We still cannot predict exactly when the MRCA lived, but the law of averages will help us decide if two men could have a common ancestor within a certain time frame. Too many mismatches will rule out the possibility, while a good match will support the theory. If you care about the details, we will cover some statistical principles in Chapter 10, Interpreting and Sharing Results. But for now, you have all you will need to know if you grasp one fact: Y chromosome tests cannot prove that you share a *particular* common ancestor with another person, only that you share a common ancestor at some point.

Define the Purpose

Because DNA testing for genealogical purposes is still—at least in relative terms—a shiny, new toy, there are those who want it just for that reason. They don't really understand it, but they want to be the first on the block to have it, so they dash out, and send for a test kit. We can relate to this because both of us are consistently among the first in line when a new DNA test becomes available. We like to take it and dissect the results, trying to better understand how it works and its potential uses. But for most people, it doesn't really make sense to take these tests for the heck of it.

Anyone who rushes out and takes a test blindly hoping to learn something is likely to find himself disappointed when he gets his report with its jumble of numbers. Such people are apt to wind up asking "Now what?"—or worse, "So what?" So please don't make the mistake of testing in the hope of stumbling onto something interesting! In the future, when large numbers of people have been tested and accessible DNA databases are exploding with samples, the odds will improve that a random person could get tested and discover something interesting, such as a surprise match with a stranger. But we're not quite there yet.

You're much more likely to find satisfaction if you join (or start) a surname project. We'll supply more details on how to do that in Chapter 7, Joining or Running a Project.

Do You Really Want to Know?

Before we move on to possible reasons for testing, we would be remiss if we didn't offer a word of caution: Be sure you really want to know! As you read through all the examples and start formulating possible scenarios to test in your own family tree, ask yourself how you (and others) would feel if the result was *not* what you were expecting.

What if you launch a study for your surname and discover that you're a genealogical orphan—not related to anyone else with your name? It happened to one of us! What if you innocently ask Granddad to get tested, only to discover that he doesn't match his own brother—that is, encounter a non-paternity situation? What if you're absolutely convinced of the accuracy of your research, and the test results contradict your years of research? How will you feel if any of your long-held notions are refuted by the testing? If you're like most of us poking into our roots, you're a truth-seeker, so this gentle disclaimer won't deter you. In fact, the prospect of a surprise of some sort is exactly what motivates some people to become involved in DNA testing, but it's useful to recognize that others may not share your enthusiasm for unexpected revelations!

Non-Paternity

Non-paternity is a catch-all term for situations where the Y chromosome is unlinked from the surname. Common causes are formal and informal adoptions, infidelity, illegitimacy with the child assuming the mother's surname, and aliases or other intentional name changes. The non-paternity rate will vary widely, depending on the time period and cultural factors.

Bryan Sykes, a geneticist at Oxford University, studied 48 men with his own surname from the Yorkshire area. The name was first recorded in the 13th century. He found that more than half of the Sykeses showed the same (or very similar) DNA pattern, while it did not occur in a control sample of non-Sykes men living in the same neighborhood. The cluster carried a signal from a single common ancestor.

The Sykes name means "boundary ditch." Could the name have been adopted by other men living near a ditch? The remaining results were widely scattered, so there was no evidence for another founder. Non-paternity events in any generation after the MRCA could cause the disparate results. Sykes calculated that a rate of 1.3 percent per generation would account for his findings. That's lower than many people in today's world might predict, but it has a cumulative effect, adding up over the generations.

Many surname projects have less ambitious goals than reaching the 13th century. Is it reasonable to identify a common ancestor from colonial times, perhaps 12 generations back? Using the 1.3 percent non-paternity rate, more than 85 percent of the nominal descendants will carry a recognizable signature. The remaining 15 percent may possess the founder's name but someone else's Y-DNA.

Y-Chromosome Objectives

If DNA testing cannot pinpoint a particular ancestor, what kinds of questions can it answer? We've gathered a number of examples from experienced project managers to start a brainstorming session for you. Grab your pedigree chart, and see if you can find parallel situations in your ancestral lines. You'll probably look first at the top line on the chart, since that is the one that follows your surname. Don't feel confined to the straight paternal line, though. Some of the "inside" lines may attract your attention too. You won't be able to use your own DNA, but if you have some male cousins who carry the surnames associated with these other branches of your family tree, you can borrow them for the actual test. If you don't have even a single candidate on hand, you may be spurred to start a treasure hunt. (See Chapter 8, Finding Prospects.)

(continued on page 44)

Figure 3-2: Typical Y-chromosome testing objectives

Typical Objective	Theory Being Tested
Standard Surname: Finding connections among anyone named X	We believe that various clusters of those named Hull share common origins. If we build a database of the assorted Hull lines and group them by their DNA signatures, those within a given cluster can compare lineages in the hope of establishing connections. And those less advanced in their research can benefit from learning where to focus their future efforts.
Suspected Surname Connection: Proving that families with the same surname are related	We think that Edward Perkins of New Haven Colony was related to John Perkins of Massachusetts and Abraham Perkins of New Hampshire. There's been a 200-year debate about whether two Hoyts—John and Simon—who were in Massachusetts in the 1630s were brothers, but I suspect they weren't.
Family Elimination: Ruling out others with the same name as being related	Since the name Smith is so common in certain southern states, we can make our research a little easier by at least proving which families are definitely *not* related to each other. Most records in Sevier County, Tennessee, were lost, so it's been challenging to sort through the assorted Laymon/Layman families. Many believe they're all related, but my research indicates that there were probably two families with these names.
Name Change/Variation/Alias: Finding proof of surname modifications	We suspect that our name was changed from Schnut to Schmidt. We believe that Flatt, Flath, Van Vliet, Van Der Fleet, and Vandervleet may be variations of the original Dutch name Van Der Vliet. Because the (Mac)Gregor name was illegal at one time, some were forced to adopt aliases. One of these aliases may have been Stirling.
Rare Surname: Proving that everyone named Y is related	Because it's such an unusual name, and everyone with it comes from the same village, we believe that all Smolenyaks share a common ancestor. The Probasco name is quite rare, and some with this name can trace their roots to a man who immigrated to New Netherland in 1654 with the Dutch. We think that it's possible that all Probascos share a common ancestor, possibly this immigrant.

Typical Objective	Theory Being Tested
Uncertain Paternity: Uncovering connections hidden by adoption, illegitimacy, and other circumstances	We think that Grandfather Howery obtained his name from his stepfather and was not really a Howery by birth. Great-Grandma worked as a maid for a wealthy gentleman named X. Although she married Y, family lore suggests that the real father of her oldest son was X.
Famous Roots: Discovering if family tales of famous relatives are true	Thomas Jefferson fathered the children of Sally Hemings. We've always been told that we're descended from Benjamin Franklin, but have our doubts. One of my great-great-great-grandmothers was a courtesan and favorite of King George III, and at least one of her sons was probably his child.
Research Verification: Putting traditional genealogical research to the test	We want to validate our written documentation about the Herrick family in North America. I am confident in my Stephens research of the past 20 years, but want scientific confirmation through DNA testing. I'm not confident in my Schreiner research. The evidence is inconsistent, and I might be indulging in some wishful thinking.
Crossing the Pond: Trying to discover geographic origins in another continent or country	We want to try to connect the Mireaults of North America to those in France. The Jarman name is widespread in the U.K., but by testing British Jarmans, we can pinpoint geographic origins for particular American families. I have reached a dead end in the paper trail in Ireland, and since there are only a few Kincaid families in the county my family emigrated from, testing some Kincaids who still live there today will tell me which family I belong to. Although we're African-American, family lore says that our male progenitor was a slave owner of European heritage. We'd like to see if this is true.

You've probably realized that there's some overlap among various objectives. For instance, the well-known Jefferson-Hemings situation is listed as a famous roots case but could also be considered an example of uncertain parentage. In fact, there's frequently an element of uncertain parentage in famous roots scenarios. Also, it's very possible for a single project to have multiple objectives as in this real-life study:

> *"We have several objectives including (1) validating our written documentation about the Herrick family in North America, (2) proving a connection between 'unclassified' Herricks, and (3) finding possible connections to Herricks in the U.K."*

And it is inevitable that new purposes will emerge as we all become more familiar with the technology and what it can and can't do. For instance, a small but growing number of people are extending their initial surname projects to include others who lived in the same area (usually restricted to a defined geographic region) as their ancestors. Such studies already exist for the Shetland Islands, Puerto Rico, the village of Osturna in Slovakia, and the village-clusters of Englisberg, Obermuhlern, Niedermuhlern, and Zimmerwald in Berne, Switzerland, and Gioiosa Ionica and Martone in Calabria, Italy, among others. In all cases, it is believed that more can be learned about the deep ancestry of an area by studying multiple names. And as some have already discovered, sometimes the answers to our personal history mysteries reside in the DNA of our ancestors' neighbors!

Do You Have a Match?

For most theories, the hoped-for outcome is a match. A match can prove, for instance, that families tracing their roots to colonial Massachusetts and Virginia actually share a common ancestor. Or it can testify to the accuracy of your research or establish once and for all that you really are related to someone

What Is a Match?

Your DNA report will list the names of the markers (e.g. DYS390) and the results (e.g. 24, which is the number of repeats). It's often convenient to string the numbers together in a compact fashion: 14–12–24–11–13–13. That makes it easy to line up a whole set of results and spot the similarities and differences, but different companies use different STRs or report them in a different order, so you must be sure you are comparing the same markers.

Your *haplotype* is simply your complete set of results on whatever markers were tested. The most common haplotype in Europe is called the Atlantic Modal Haplotype (AMH), because it is frequently found in countries near the Atlantic Ocean. The markers used to define the AMH are DYS19, DYS388, DYS390, DYS391, DYS392, and DYS393. In the four samples below, A and C both have the AMH, and they are perfect matches with each other, but B has a mismatch on the third marker. D only matches the others on two markers.

A. 14–12–24–11–13–13

B. 14–12–23–11–13–13

C. 14–12–24–11–13–13

D. 14–14–22–10–12–13

Since a mutation can occur at any time, B may be closely related to A and C (note the change from 24 to 23). However, D has so many mismatches that it's clear he does not share a recent common ancestor with the other three.

The six markers listed above are classical markers, which have been used in almost all research studies. All the genealogical testing companies offer these six. Then each adds its own selection from more recently discovered markers. There are currently several dozen markers that have been studied in different populations around the world.

famous. But there are a few exceptions. For instance, in the elimination scenario, the desired result is a mismatch. The fewer matches, the fewer Smith family records you'll have to wade through in your paper-trail research!

Whatever you might hope the outcome will be, the measure of a good theory is simply that the testing answers your question—or at least moves you closer to an answer. Haphazard testing in the hope of matching just anyone isn't apt to be productive at this time, but we hope you'll agree, after browsing through all the examples we've shared, that there are plenty of worthy Y-chromosome testing objectives. Let's turn now to some real world examples to get a better sense of what you can expect.

Very Common . . .

In the world of genealogy, it's often debated which is worse—to bear a surname that's wildly common and forces you to sift through countless candidates for your ancestors, or to have one of those rare names that everyone stumbles over and misspells, resulting in forebears who are hidden under distorted versions you never could have imagined. Fortunately, genetealogy can help with both situations.

Some with common names have turned to genetic genealogy to help narrow the field. For instance, Debbie Harper and Wayne Bates jointly coordinate the Southern Smith/Smyth/Smythe/Schmidt DNA Reconstruction Project designed to "sort out the various Smith lines that were at one time in the southern states such as Virginia, Kentucky, South Carolina, North Carolina, Georgia, and Tennessee."

As Debbie explains, "The sheer number of Smiths makes it extremely difficult to figure out which Smiths one might be descended from, but the thought was that DNA samples could more clearly differentiate branches." By sharing their results, those pursuing their Smith lines will have a better chance of finding distant cousins as well as the ability to rule out certain branches, thereby saving considerable research effort that would

Haplotype Diversity

How often will two random Smiths match each other just by accident?

Just as surnames can be very common or very rare, haplotypes are found in different frequencies. In the database at www.yhrd.org, which has more than 24,000 records tested at nine markers, the single most frequent haplotype occurs in less than 3 percent of the population, so even that could not be called common in the absolute sense. Many haplotypes occur just once—more than 40 percent of the records, in fact. Every time a new set of data is added to the database, novel haplotypes are discovered.

Haplotype diversity can be quantified. The chance that two men chosen at random will match each other on all nine markers is less than two in a thousand. You can rule out a lot of false trails that way, and if two Smiths match, it's probably not just a coincidence.

Adding more markers increases the diversity: Some of the men who match on nine markers will differ on a 10th marker.

have eventually led to dead ends. Although the project is focused on southern states, Debbie says that they hope to also compare results with another Smith project for Massachusetts, Connecticut, and New York.

Other names that are being researched in a similar manner include Johnson, Clark, Brown, Rose, Graves, Walker, Rice, Bassett, and Stewart.

. . . Or Very Rare

The flip side of this, of course, is coping with a rare name. You know you fall into this category if one of your family traits is checking for your name in hotel room phone books whenever you travel. Many of us, including one of the authors, grow up suspecting that everyone with our name must be related somehow. But is that really true? As you'll see from the following example, DNA testing finally gives us a way to find out.

Through three decades of research, it was discovered that all Smolenyaks in the world (now scattered from Ukraine to Antarctica) hail from the village of Osturna, located in present-day Slovakia. Osturna records dating back to the 1700s revealed that there were four Smolenyak families who had been situated—with remarkable stability—in houses 88, 96, 103, and 135 for centuries. While it seemed reasonable to assume a common ancestor, the paper trail had failed to yield any such proof.

Conveniently, at least one man from each of the households had emigrated to the United States and had living male descendants, so a DNA project was launched with a representative from each line. The initial hypothesis was not that all four would match, but that three of the four would. It was theorized that the fourth line would not match due to a blended family situation around 1800. A widow and widower married, resulting in offspring that were his, hers, and theirs. Over the ensuing decades, children from all three unions casually bounced back and forth between the surnames of their respective birth fathers and stepfathers, and it appeared that a few who ultimately sported the Smolenyak name were not Smolenyaks by birth. In short, it was believed that one line of Smolenyaks might really be Vaneckos.

When the results came back, everyone was stunned to learn that *none* of the four matched! So unexpected were the findings that a second round of tests was done, this time including a Vanecko participant to explore the wrong-surname notion on the one branch. The second batch confirmed the first, with the added insight that the fourth cluster were indeed Vaneckos-in-disguise.

While this abrupt unlinking of Smolenyak lines was disappointing in a sense, all the Smolenyaks were relieved to have saved themselves decades of trying to prove an apparently false belief, and for the one branch, it was especially interesting to learn what their name would have been, were it not for a step-family situation about two centuries ago.

Recognizing that the DNA of our ancestors' neighbors could possibly answer more questions, the study was then opened to anyone with Osturna roots. To date, participants with 13 of the village's 50 or so surnames have been tested. The aim is to eventually get at least two samples for every surname native to Osturna. So far, some of the results have confirmed connections (all those with the most common name in the village apparently do share a common ancestor) and hinted of unexpected ones (one of the Smolenyak lines recently matched a participant with the name Homza). Every fifth test or so offers a surprising result, and it is these unexpected findings—the prospect of learning what the paper trail can never reveal—that keep the study fascinating to its participants.

Why Two Samples?

Why did we specify *at least* two samples in the preceding example? Test results for just one person are like "the sound of one hand clapping." That one person can't reveal anything about other people with the same surname. Two samples (if they match) will demonstrate the haplotype of their Most Recent Common Ancestor.

The more distantly related the two people are, the deeper the chain of evidence. If you refer back to Figure 3-1, you'll see that E's haplotype is evident if I and J match, but B's haplotype is not proven. A match between I and K will confirm B's haplotype (since B is their closest common ancestor), and a match between I and M will lead the way to A's haplotype. *We do not need a direct test of A's DNA!* It is preserved in the Y chromosome of his descendants. If you've been picturing an army of genealogists prowling through cemeteries at the stroke of midnight, you'll be relieved to know that exhumations are not required.

If the two samples disagree on one marker, then one line had a mutation, but it's impossible to tell which one. A third sample will be needed for a tiebreaker, but you'll end up with a bonus: a mutation that marks one branch of the descendancy chart.

If the two samples are markedly different, the descendancy

chart is questionable. Then it's time to review the paper trail. You may spot some weak links, connections that seemed plausible based on circumstantial evidence. But if you can't poke any holes in your own theory, then you'll need to decide whether to obtain samples from other men from the same line to rule out the possibility of a non-paternity. Your approach may have an impact on people who are not even participating in the surname project.

Uncertain Paternity

Situations involving uncertain paternity have posed a challenge to many genealogists. Whether we know of or have heard whispered family tales of adoption or illegitimacy, the information usually leaves us stranded in "what now?" territory. Justin Howery is one researcher who's used DNA testing to try to knock down this particular brick wall.

According to Justin, "We Howerys (Howrys, Hauris, Haurys, Howreys) supposedly all descend from a single ancestor who lived circa 1400 in the Swiss village of Beromünster," but some present-day researchers had their doubts. In fact, family lore in Justin's branch held that they weren't Howerys at all and had "just acquired a stepfather's surname somewhere along the line." More specifically, he had been led to believe that one of his paternal grandfathers (perhaps his third great-grandfather) was adopted and that his birth name may have been Hamilton.

The research hadn't turned up any evidence of a Hamilton connection, but Justin thought that the tradition probably had some truth to it, as these stories so often do. So when assorted participants on a mailing list for the Howery surname began discussing Y-DNA testing, he was enthusiastic about joining to at least prove the accidental-Howery aspect of the family tale. Unlike most participants who anticipate a match, Justin hypothesized that he would *not* match anyone else in the study. As he explains, "I was expecting a dramatic disconnect between my results and those of my test partner. Then with a heavy sigh, I would turn from Howery research and start looking for that elusive stepfather."

His chances to prove or disprove his theory seemed slim as only one other Howery followed through with the first round of testing, but Justin was astonished to discover that the two of them were a perfect match. The family tale, whatever its origins, was apparently wrong, casting an unfounded shadow of doubt over decades of research. As he puts it, "I now have the first real evidence that my descent is really through the Howerys. The test results of just two guys—albeit the right two guys—dramatically swept aside a lot of meaningless and irrelevant what-ifs. Years of questions were put to rest in just 6 weeks by DNA testing, and now I won't waste any more time looking for a Hamilton connection that doesn't exist."

Geographic Origins

Another popular reason for testing is to learn more about one's geographic origins, an objective that's more challenging for some of us than others. Like many Irish-Americans, Clarke Glennon had run into a brick wall in his research. Historical events have conspired in Ireland to make it extremely difficult to get beyond the 1800s, particularly for those who hail from Catholic families. Clarke could go as far back as 1845 to his great-great-grandparents in County Galway, but he still had plenty of questions.

> *In Ireland, the Glennon name historically is well-known in Counties Galway, Roscommon, Longford, Westmeath, Offaly, Leix, Meath, Kilkenny, and nearby. Do all descend from one man, a Glennon Adam? Or are there unique Glennon families in each of these locations? Are the McLennans and McGlennons of Scotland and Northern Ireland related to them? Irish surname experts link names such as Glynn, Leonard, and Lennon to the same origins. How correct is this?*

When genetic genealogy became available, Clarke quickly realized its potential and asked fellow Glennons as well as those with similar names to participate in a surname project. In fact, the questions above are posted in his online invitation to join his DNA project, where he goes on to say, "Join the Glennon study, and see how your family Y-DNA compares with others in the study." And it wasn't long after its launch that Clarke delivered on his promise.

Among the first 10 participants were Brian Glennon and Thomas Glennon, both of Massachusetts. Business acquaintances with a shared interest in their heritage, they had often jokingly referred to each other as cousin, but an examination of their family trees revealed no connection. Still, they were both curious enough to join Clarke's study as soon as they learned about it. Much to everyone's surprise, the testing revealed a perfect match, indicating that Brian and Tom did indeed share a common ancestor. Now when they call each other cousin, it's with the certain knowledge that they truly are.

Clarke's project is now more than 2 years old and, at last count, had 53 participants—enough to allow him to start answering some of his questions, especially about his origins in Ireland. In a recent update, he reported, "The results so far show that all Glennons who live in or trace their origins to either County Roscommon or County Galway are descendants of one man. All are distantly related. On the other hand, in County Westmeath, there are at least three different, unrelated groups of Glennons whose origin is in different parts of the county. There are also Glennons from County Kilkenny, County Cork, and County Cavan who are probably not related to the others."

There's also been an interesting recent twist involving a name change. Clarke persuaded a man named Lennan, whose family's paper trail went back to before 1850 in the Dublin area, to join the project. He was surprised to match 25 for 25 with a man named Glennan, whose roots were 50 miles west of that area, in County Meath. As still other Glennons joined the project, both men became part of a cluster of Glennons whose origin is in the region where the counties of Westmeath, Meath, and Offaly come together. This steered the Lennan man's research to a different place and even to a slightly different surname than he would have pursued otherwise.

Clarke believes that as the results accumulate, the prospect of a random Glennon testing and quickly learning where to focus his research efforts in Ireland will only improve, especially now that Glennons from 12 counties in Ireland are now represented in his study. Moreover, those who match him will form a ready-made team of allies in their shared genealogical quest.

Name Changes

For some of us, our own name may be the source of confusion. If that's the case with you, it's possible that genetealogy can provide a little clarity or insight. Our ancestors frequently tried their best to blend in when they hopped continents, and one of the most common tactics was altering or changing their name to something that didn't stick out as quite so foreign in their new environment. Many Schmidts of German origin, for instance, have changed their name to Smith, so researcher Tom Schmidt of New York would have been smart to extend his ancestral research to the name of Smith. What he encountered, however, was somewhat less expected.

After 15 years, Tom's research led him to the conclusion that his Schmidts were actually descended from a German family named Schnut. It appeared that his ancestors had indeed changed their name to something more common in their new homeland, but they weren't quite willing to shed their cultural heritage. Still, Tom wanted to be 100 percent certain of what the paper trail seemed to be telling him, so he located a few Schnuts still living in Germany today. Fortunately, they were genealogists themselves and had a basic understanding of DNA testing, so they readily agreed. The result? A match confirming Tom's name change hypothesis and the removal of any lingering uncertainty.

Tom's situation is far from unique, although the reasons behind the name changes vary. In fact, one of Y-DNA's fastest growing purposes is to substantiate suspected name changes or uncover previously unknown ones. Here are a few examples:

- Jesse Moore's family passed down a story through the generations that their name had been changed from Kirkpatrick at the time of the Revolutionary War. When he was tested, Jesse discovered that he was just a one-step mutation away from a man named Kirkpatrick. Since the odds of such a near match occurring accidentally are exceedingly slim, the family tale has been largely corroborated, but additional testing of Kirkpatricks residing in the same area as Jesse's ancestors can provide further confirmation.
- Nancy Custer spent years researching a hushed family tradition that her great-great-grandfather, Harvey Kelley of Cumberland County, Tennessee, began life near Walhalla, South Carolina, with the surname Dorsey, and she succeeded in building a strong circumstantial case that the rumor was true. When DNA testing became available, she had descendants of Harvey Kelley and a proven Walhalla-origin, Dorsey male tested. They matched 25 for 25 markers, verifying the patchy paper trail. The reason for the name change remains a mystery, but Nancy's genealogical brick wall has finally been knocked down.
- According to Richard McGregor, tradition suggested that all who have the (Mac)Gregor name would be related because of the severity of legal proscription on the name (imposed in an attempt to dampen Scottish nationalism) in the early 17th century. Another result of the banning of the name was that members of Clan Gregor have many aliases today, a notion that was put to the test through the use of DNA. As expected, a male with the name of Stirling, one of the suspected aliases, matched others in the clan.

Mystery Matches

Some people are actively searching for testing candidates with different surnames. They are delighted if a match confirms their theory of a name change. What if, completely out of the blue, your DNA matches someone with a different surname? Non-paternity in one line or another is always a possibility, but the history of surnames can be revealing too.

Sometimes the names are obvious sound-alikes, Laymon, Lemon, Lehman, and LeMond, for example. Other times, a German or French name may be Anglicized: Zimmerman may turn into Carpenter, or Beauchamp may become Fairfield. Some countries have used "patronymics," where a son uses his father's given name, so it's not surprising if a Williamson matches a Johnson and a Davidson.

There's still another possibility. Your Most Recent Common Ancestor may not be very recent at all. He may have lived before surnames came into use. Different descendants may have adopted names derived from their occupation (Baker) or personal traits (Whitehead) or place of residence (London). It may never be possible to find the connection, but if you learn that your ancestors lived in the same locality at the same time, you may find yourself doing some more detective work.

When the Paper Trail Is Confusing

One of the authors enlisted a male cousin to solve a longstanding genealogical mystery about her great-grandfather. A biographical sketch of an ancestor often begins in a straightforward way with a name, a date, and a place, as in "William Henry Shrinard was born September 17, 1850 in Cass County, Michigan." These items about my grandmother's father were gleaned from several traditional genealogical sources, and the birth date was even carved in stone, as the saying goes. But there were elements of doubt in all three. William Henry, who went by Henry, was orphaned at a young age. In 1870, he traveled with another family from Michigan to Barry County, Missouri, where his name appeared on his marriage license and census records as Shrinard. A search of Michigan court records turned up guardian papers for Henry Shrinard, son of John Shrinard deceased. That clue was an immediate dead end, as the name Shrinard did not occur anywhere else in that county—nor indeed, even in the 1850 census records for the entire country! Shrinard must have been the way the court interpreted the young boy's pronunciation of his name.

There was, however, a marriage record in 1852 for a John Shreiner, aged 35, to a young widow, Mary Ann White, who had a son of her own. Could John *Shreiner* and John *Shrinard* be the same man?

John and Mary Ann had a son together, by the name of Leroy. If John Shrinard was indeed John Shreiner, Leroy would have been Henry's younger half-brother. John apparently died before the 1860 census, because Mary Ann, with her son from her first marriage and Leroy from her second marriage (but without her stepson, Henry), was married to another man by then. That hybrid family moved to Washington about the time that Henry moved to Missouri.

It took fifteen years of research, but one of Leroy's descendants (who spelled the name Shriner) was located and kindly agreed to take a DNA test. His Y-chromosome results showed an exact match to those of my cousin, and the haplotype proved so rare that it was not found in any of the existing databases. "William Henry Shrinard" had apparently been born Henry Shreiner, and a family that had been separated for more than 140 years was finally reunited!

But the tale continues. Henry's father, John Shreiner, was almost as elusive as the nonexistent John Shrinard. There were no other records in Cass County, Michigan, for anyone by that name. But wait! The 1850 census for Elkhart County, Indiana, just across the state line, listed a John Shreiner, living next door to his brother George Shreiner. John's age matched the marriage record in Indiana, and he had a son named Henry, age 8 months. Yet this Henry's birth date and birthplace would conflict with other records gathered in Barry County, Missouri. Would it be jumping to conclusions to assert that John of Michigan and John of Indiana were the same person?

The Shreiners of Indiana can trace their ancestry to Lancaster County, Pennsylvania. The rather unusual spelling was adopted by the immigrant Hans Adam Shreiner, who dropped the "c" from Schreiner when he came to America in 1744. It's thought that many who use this spelling descend from Hans Adam, and the highest prevalence of the spelling Shreiner is found in Lancaster County even today.

Could a connection be made by testing a Shreiner—any Shreiner—from Lancaster County? A volunteer stepped forward, and the results again came back with a perfect match. The DNA had overcome the gaps and contradictions in the paper trail. It linked families living in the 19th century, backward and forward, to families in the 18th century and the 21st century.

Why Y?

Surname projects are the most common application for genetealogy. By the time you read this, there will undoubtedly be more than a thousand of them, both big and small. We're the beneficiaries of a fortuitous combination of many biological and historical circumstances. The Y chromosome follows a well-defined path down the straight paternal line, which oh-so-conveniently tracks the surname in many cultures. The details on the Y chromosome are not shuffled and blurred by mixing with other chromosomes. A stable surname makes it "relatively" easy (pun intended!) to follow the genealogical paper trail and locate recruits for a project. We can use the very same markers developed by research scientists for our own humble goals. And the mutation rate turns out to be "just right."

Some critics dismiss the value of genetealogy because it can not prove descent from a particular ancestor. Proof is a strong word. DNA testing is a *novel, objective,* and *independent* form of evidence, which can support, validate, confirm, reinforce, substantiate, and corroborate other items of evidence you have acquired for your theory. (Or, to your temporary dismay, it can negate, refute, contradict, and discredit your current hypothesis.) DNA testing does not stand alone. It goes hand in hand with traditional genealogical research. And that's a potent combination! But that's not the end of the story. There are ways to trace maternal ancestry too.

4

Maternal Legacy

Mitochondrial DNA

We turn now to mitochondrial DNA, in some ways a parallel to the Y chromosome. Just as the Y chromosome follows the straight paternal line, mitochondrial DNA tracks the straight maternal line. It is the smallest of the DNA molecules, but its importance far outweighs its size.

A Different Kind of DNA

When we speak casually of DNA, most often we are thinking about the DNA that is found in the 46 chromosomes, inside the nucleus of the cell. The remainder of the cell contents—the cytoplasm—harbors DNA as well, inside small structures called mitochondria (my-toe-CON-dree-uh, the plural form of mitochondrion). About the size of bacteria, hundreds or even thousands of mitochondria are crammed into one cell. In fact, the very size is a clue to the origins. Sometime in the distant past, one primitive life form stumbled inside another and stayed put.

All organisms with a nucleus, from single yeast cells to giant whales, have these permanent guests, the mitochondria.

The mitochondria are considerate of their hosts, however. They pay the energy bill by breaking down complex food molecules in gradual steps instead of burning it in a blaze. (*Star Trek* fans may remember an episode where mitochondria were likened to the warp core of the cell.) If the cell requires lots of energy, the mitochondria reproduce by splitting into two. Indeed, one effect of exercise training is to ramp up the number of mitochondria in muscle cells.

Passing the Torch to the Next Generation

When it comes time to create an egg and a sperm, the two sexes have different strategies.

The plumped up, nurturing egg is much larger, and it may contain as many as 100,000 mitochondria. The swimming sperm, by contrast, is a streamlined package, with nuclear chromosomes packed into the head and a few mitochondria positioned around the tail to fuel the journey to the egg. When the sperm penetrates the egg, the tail drops off. Not only are the mitochondria from the sperm outnumbered by 1,000 to 1, the egg actively disposes of any mitochondrion that slips inside.

This lopsided mode of inheritance has profound implications. If the mitochondria all come from the egg, they therefore come from the mother. Where did the mother obtain her mitochondria? From her mother, who inherited them from her mother, who inherited them from her mother, back for countless generations. That follows the bottom line on the pedigree chart (as seen back in Figure 1.1). Note that this is different than the mode of inheritance for the X chromosome, which was covered in detail in Chapter 2, Genetic Essentials.

Mitochondrial Eve

In Chapter 3, Male Bonding, we saw how the paternal ancestry of all men living today converges on one man, Y-Adam. Can everyone living today trace their straight maternal ancestry back to one woman? The answer is yes, and she is dubbed mitochondrial Eve. It's the same scenario. Each generation of daughters coalesces to a smaller number of mothers until there is only one remaining. Conversely, the mitochondrial DNA of Eve's contemporaries has disappeared forever. See Figure 4-1 for a small-scale illustration of how that might happen.

Like Y-Adam, mitochondrial Eve was born somewhere in Africa, but she did not live at the same time as him. Her date of birth is uncertain, but most estimates fall within a range of 120,000 to 200,000 years ago, long before Y-Adam. Adam never met Eve!

Mitochondrial Eve was not the first woman, nor was she the only woman living at the time. But the mitochondrial DNA of everyone living today harkens back to that one person. The subtle individual differences we see now are the result of mutations in various branches of Eve's descendants, which have slowly accumulated in the intervening years.

Figure 4-1: Path of descent for A's mitochondrial DNA (mtDNA)

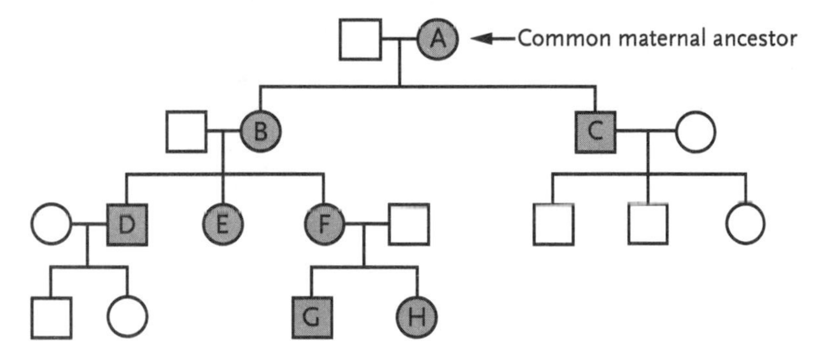

All the shaded figures (the males in squares and the females in circles) inherit their mitochondrial DNA from A. The children of the males C and D receive their mtDNA from their respective mothers. If A's great-granddaughter H has no daughters, A's mtDNA will disappear forever, and she can never be mitochondrial Eve, although she may have many descendants who carry some of her nuclear DNA.

References Supplied upon Request

Literally speaking, your results from a mitochondrial DNA test are simply a long string of letters, TATTGACTCA . . . Depending on the specific test you order, the length of the string will range from 340 to 1,100 bases. If I read my report to you, and you read your report to me, we'd probably both fall asleep before we could decide if we were the same or different. For the sake of convenience, everyone is compared with a single reference point, and only the differences are reported. Typically, there will only be a small number of these genetic differences or polymorphisms.

So who's this reference point? It all started in 1981 when S. Anderson at Cambridge University, using a sample from the placenta of one anonymous local woman, first sequenced mitochondrial DNA (mtDNA for short, pronounced em-tee DNA). Anderson published the whole sequence—that is, the exact order of the bases A, C, T, and G—in *Nature*. The molecule is 16,569 bases long (a mere tidbit compared with the three billion bases of the nuclear chromosomes), and the sequence fits on three pages. The result has been variously called the Anderson sequence or the Cambridge Reference Sequence (CRS).

There is nothing special about the Cambridge Reference Sequence, except that it is the first time mtDNA was ever sequenced. The woman donating the placental sample was not chosen for any reason, except that she happened to deliver a baby in a hospital near the researchers at the time they were ready to get to work. The CRS is not the same as mitochondrial Eve, or even the sequence that is closest to Eve. It is not the "best" version of mitochondrial DNA, either, but by chance, it is one of the more common patterns found in Europe.

Everyone's a Little Bit Hyper

Mitochondrial DNA forms a closed circle, so the numbering could start anywhere, but by convention, base #1 is in the center of the "D-loop." The D stands for displacement, a little section that bulges out when the mtDNA begins to make a copy of itself. This region is sometimes called the control region (as opposed to the coding region, which we'll discuss in a moment), because it seems to initiate the replication. Most often this region is called the Hypervariable Region, or HVR. "Hyper" may sound a little scary, since it is a word prefix for excessive, or at least more than normal. What's normal for mtDNA?

The genes in the coding region must be carefully conserved. The proteins have been optimized over millions of years, and mutations could have a major impact on their function. If it ain't broke, don't fix it! However, the D-loop appears to act as a spacer. It doesn't seem to matter much what base goes where, as long as there's room to spread apart when replication begins. That means that mutations can pile up without causing harm. In fact, no diseases have been found that are caused by changes in the HVR.

As more and more samples were studied and compared with the CRS, it became clear that polymorphisms were much more likely to show up within the range covered by the D-loop. "Hyper" is thus a relative word, comparing one part of the mtDNA molecule with another. In absolute terms, the mutation rate is actually very low. Most people have only a handful of differences compared with the CRS (or to any other person, for that matter).

Test Reports

Every laboratory has customized its offerings with various add-on features, but they all have a common core: They are sequencing several hundred bases in the Hypervariable Region. Why not the whole thing? After all, it's only 16,569 bases. Current technology can only handle a few hundred bases at a time. It would take 2 dozen to 3 dozen separate reactions to assemble a composite sequence for the whole molecule. Fortunately, the HVR is the most informative section, so concentrating efforts there offers the most bang for the buck.

Some companies may list your actual sequence, while others will simply point out where you differ (if at all) from the CRS.

Although your report may list your "mutations," it's technically more correct to say differences or polymorphisms. The CRS may have the mutation compared with mitochondrial Eve, while you may have preserved Eve's original version.

Your report might read 16224C 16311C. That means that you have the base C at positions #16224 and #16311, where the CRS has a different base. It doesn't matter what the original base is (it happens to be T in both cases), but if you wish, you can view the entire reference sequence at www.mitomap.org/mitomap/mitomapRCRS.txt. (The R in RCRS stands for Revised. In 1999, the original sample was reanalyzed with modern techniques. Very few changes were made, none in the Hypervariable Region. Some laboratories may compare your results with the CRS and others with the RCRS, but the end result is the same.)

Your complete set of polymorphisms is your haplotype, which you might remember is the complete set of results from tests on a single chromosome, one that was inherited from one parent. Your mtDNA polymorphisms come as a package deal, not as a potpourri of contributions from different ancestors.

So, would you be jumping up and down in excitement to learn that your haplotype is 16224C 16311C? That's easier to read and remember than a long string of letters, but only marginally more interesting. The plot thickens when you can compare your results with other people, not just the CRS.

Compared with Whom?

At the time of this writing, the haplotype 16224C 16311C has been found in 180 out of several thousand customers of Oxford Ancestors (www.oxfordancestors.com), one of the genealogical testing companies. These 180 all have a common maternal ancestor, somewhere back in time. When, you ask? Therein lies the rub. It could be recent, or it could be thousands of years ago. In fact, this very haplotype was found in the Iceman, the 5,000-year-old mummy recently recovered from the edge of an ice field in Italy! The haplotype is also characteristic of one of the "The Seven Daughters of Eve" that Bryan Sykes describes in his book.

This vagueness about the time of the Most Recent Common Ancestor (MRCA) is not unique to mtDNA testing. As we saw in the previous chapter, Y-chromosome tests cannot be completely specific about the time frame either. But the problem is exacerbated for mtDNA testing. Mitochondrial DNA changes more slowly than the Y chromosome, and the time frame broadens considerably. Random matches in a database are less likely to be fruitful.

We have already stressed the importance of a hypothesis for Y-chromosome testing. The same principle applies in spades to mtDNA testing. Comparisons between two specific people can reveal the presence of a connection, and mtDNA testing offers a compensating advantage: It is not limited to the living. After death, bacterial action cuts DNA into smaller and smaller fragments, sometimes too short to test. But there are thousands of copies of mitochondrial DNA per cell, and fragments only a few hundred bases long can be highly informative.

But, you protest, this doesn't apply to me. I can't just go around digging up my ancestors to satisfy my own curiosity! True, but you don't need to. There are plenty of genealogical puzzles that can be addressed using the mtDNA of the living, as one of us first realized more than a decade ago.

Proof of Principle

The potential of mtDNA testing was first explored in a 1992 article in the New England Historic Genealogical Society publication *Nexus,* titled "Mitochondrial DNA: A Genetic and Genealogical Study." Thomas H. Roderick, Mary-Claire King, and Robert Charles Anderson proposed collecting long matrilineal lines for an academic study. The objectives were threefold: for genealogists to document and verify their matrilineal ancestry, for population geneticists to gain insight into the structure of early colonial populations, and for geneticists to determine more precisely the mutation rate of mtDNA.

Although the project didn't bear fruit—it was not easy to find enough suitable candidates—this article was my first expo-

(continued on page 68)

History's Mysteries

Because of its resiliency, mtDNA is involved in the resolution of many of history's mysteries that you probably happen across from time to time in newspaper articles or documentaries. If you hear of remains being disinterred in an attempt to resolve a long-standing question, chances are excellent that mtDNA was a factor. Here's a sampling of a few of the better-known cases.

Titanic Baby

A few days after the *Titanic* sank on April 15, 1912, the body of a baby boy was found by a Canadian salvage ship. When no identification was made, the crew took the boy's body to Fairview Lawn Cemetery in Halifax, where he was buried along with 120 other *Titanic* victims. His tombstone, marked "Unknown Child," has continued to draw attention for 90 years. But now he has a name: Eino Viljami Panula.

Eino was only 13 months old when he traveled with his mother and four brothers from Finland to join their father who was working in Pennsylvania. Recently, researchers located Magda Schleifer, an alleged Finnish first cousin once removed of the baby (her grandmother was a sister of Eino's mother). They compared her mtDNA with that taken from remains of the unknown child and were rewarded with a match, confirming his identity. Magda, her daughter, and her granddaughter—about the same age as Eino—all journeyed to Canada to pay their respects at the gravesite of their tiny cousin.

Vietnam Unknown Soldier

In May 1972, U.S. Air Force 1st Lieutenant Michael Blassie was shot down in Vietnam. Portions of an A-37 bomber, dog tags, a wallet, and bones were all recovered, but because so many American casualties had occurred in the vicinity, it was not possible to make a positive identification.

Twelve years later, remains from the site were interred in a ceremony at Arlington National Cemetery as the Vietnam Unknown

Soldier. Learning of advances in DNA research and believing that the remains could be Michael's, members of the Blassie family requested that they be tested. Results established with 99.9 percent certainty that the Unknown Soldier was Michael, whose family had him buried with proper honors closer to home.

Louis XVII

For more than 2 centuries, many have debated whether Louis-Charles, the son of King Louis XVI and Queen Marie-Antoinette, both guillotined during the French Revolution, died shortly after them in 1795, or whether—as some claimed—he was taken away from his prison cell and replaced by another boy. Numerous pretenders emerged after the French Revolution, each insisting he was the missing heir to the throne who had miraculously survived, but while some developed extensive followings, none were ever accepted by the sister of Louis-Charles, and the mystery lingered.

In a bizarre series of events, the heart of the alleged dauphin was pocketed by the physician who performed the boy's autopsy and eventually found its way to the basilica of St. Denis where relics of other members of French royalty have traditionally been kept.

In 2000, five samples of the heart were sent to separate labs in Belgium and Germany. Scientists compared the mtDNA extracted with that of presumed relatives of the boy king, both living and dead. Hair samples from Marie-Antoinette and her two sisters as well as tissue samples from two living descendants of the sisters were all found to be a match. A team of scientists and historians, who considered the DNA and other evidence, concluded that Louis-Charles was the young boy who died in prison.

I Remember Reading about That

These are just the proverbial "tip of the iceberg." Chances are that you've heard of similar attempts to prove through DNA testing that both Jesse James and Billy the Kid cleverly avoided their alleged demises and lived to old age while some other unfortunate fellows filled their caskets. To date, all such investiga-

(continued)

> **HISTORY'S MYSTERIES** *(continued)*
>
> tions have failed to support these appealing escape tales, but the popularity of tackling history's mysteries through DNA, and especially mtDNA, continues to grow.
>
> At present, efforts are underway to identify remains found in the U.S. Civil War submarine, the *H.L. Hunley*, and of a skeleton (suspected to be Bartholomew Gosnold) found in the walls of the original fort at Jamestown, Virginia, established in 1607. MtDNA occasionally features in modern criminal trials too. Although it cannot uniquely identify a person, it can sometimes be recovered from evidence when nuclear DNA is absent or degraded.

sure to the concept and inspired me to work intensively on my own matrilineal ancestry. I figured it was only a matter of time before mtDNA testing became commercially available. Eight long years later, my wish came true when Oxford Ancestors came online.

By that time, I had found a distant matrilineal cousin (4th cousin three times removed, to be exact). Our common ancestor was Pamela Nims, who was born in 1794 in Massachusetts. I was primed to go, so I ordered two kits. Several weeks later, the results arrived. I held my breath as I opened the envelope. Both of our sequences were highlighted in red at the same position, 16293G. We were a perfect match! The mtDNA had been passed down from Pamela Nims for eight generations in one line and five generations in the other line without changing at all. The science was not just an abstract theory to me—I had verified it for myself.

JUST A COINCIDENCE?

If people vary from each other by just a handful of differences out of the hundreds of bases in the HVR, couldn't the match confirming your research be just a coincidence? It sounds paradoxical, but there are lots of different ways to be a little bit different.

For example, H. Pfeiffer from the University of Münster, found 460 unique haplotypes using positions 16024 to 16365 in his study of 1,200 inhabitants of one German village. The most frequent haplotype, identical to the CRS, occurred in about 10 percent of the cases, but 305 haplotypes were found just once. In fact, he discovered novel haplotypes every time he added a new batch of records to his database. If more bases were included, the haplotype diversity would increase even further, and the percentage of people with exact matches would drop to still lower levels.

Mailing List Launch

This initial Nims matchmaking experience was the impetus for the founding of a mailing list devoted to the topic. That was in 2000, and it was soon clear that there were plenty of people who wanted to swap stories and study more and more about DNA testing for genealogy. Many jumped on board immediately, and message traffic increased tenfold over the first 3 years.

The archives (see lists.rootsweb.com/index/other/DNA/GENEALOGY-DNA.html) trace our learning curve, our struggles to understand and interpret results, and the appearance of every new approach in the field, not to mention creative suggestions about how to apply mtDNA testing to overcome genealogical obstacles. The following are a few examples of actual or proposed applications.

Mistaken Identity

More than a century of family tradition held that the maiden name of Lannie Walker's great-grandmother, Nancy Scott, was Perkins. He had never really stopped to question this information until a distant cousin, Gerald Whiddon, asked for verification. Apparently, Gerald had located a marriage record for John Scott and Nancy *Kirkland* on the exact date that Lannie's great-grandparents had supposedly married. Furthermore, the birthday of this newly discovered Nancy was the same as that of Lannie's great-grandmother. Could Nancy Scott's maiden name really have been Kirkland instead of Perkins?

Confronted with this perplexing situation, Lannie looked for proof of the Perkins name, but came up empty. He then decided to research the possibility that Nancy had actually been a Kirkland and quickly reached the conclusion that she had indeed been misidentified for 150 years. When he shared this new information with a fellow Scott descendant, however, she was reluctant to let go of this long-held belief. How could so many people have been wrong for so long? Not unreasonably, she wanted more proof before introducing this new information into the family history.

Lannie brainstormed for a way to provide more convincing evidence and realized that mtDNA testing might be just what he needed to settle the issue. The Nancy Kirkland he believed to have been his great-grandmother had a sister named Alcey. If "his" Nancy had been a Kirkland, her mtDNA should match that of Alcey. If they didn't match, clearly his theory was wrong. All he had to do was find appropriate descendants of both to test to make the comparison.

As luck would have it, the very cousin who was reluctant to accept the new evidence had a female-only lineage back to Nancy herself and agreed to the testing. Fortunately, a researcher Lannie had already contacted was a descendant of Alcey's. And while she didn't have the "right" mtDNA, her father—descended through a chain of daughters—did.

When the results came back, the two were a perfect match, reinforcing what the freshly examined paper trail had indicated: Nancy Scott was a Kirkland by birth. Both the Scott and Kirkland families now accept the combined traditional and genetic evidence as proof of Nancy's true identity.

Were They Sisters?

Bill Hurst's great-great-grandmother, Catherine Kelly, was born in Wythe County, Virginia, in 1807, and he hoped to learn more about her birth family. Unfortunately, two Kelly families resided in the county at the time, and he couldn't be sure which one was his. A Martha Kelly, born to John and Elizabeth Kelly about 1803, was suspected to be Catherine's sister, but proof was elusive. Could mtDNA testing be used to link the two women as sisters, thereby allowing Bill to add another generation to his pedigree?

Having come up with the idea, Bill had to find direct maternal descendants of both women. For Catherine, any of the children of his father's sisters (or their daughters' children) would qualify, and he managed to find a volunteer. For Martha, he had an obvious candidate since one of her maternal descendants had first suggested the possibility of the relationship of the two women back in 1983.

HVR1 AND HVR2

The Hypervariable Region covers approximately 1,100 bases, bracketing both sides of the arbitrary origin, so that base #16569 is right next to base #1 in this circular molecule. The region with the highest numbers—in the 16,000 range—was the first to be studied intensively and is designated HVR1. It may seem backward, but HVR2 covers the bases in the low end of the range.

The notation 315.1C means that there is an extra cytosine compared with the CRS. This is a common polymorphism, so the CRS may have the mutation, a deletion of one base compared with mitochondrial Eve. Deletions are listed with a minus sign, for example 523- 524- has been observed from time to time. Also, the CRS has the rare value at position 263G, so many reports, such as the one in this example, will list that as a "mutation."

Some testing companies test only for HVR1, some automatically test for HVR1 and HVR2, and some offer a choice. A test for HVR1 is often sufficient to place you in one of the broad groups described in the next chapter. If you have a common haplotype with HVR1, however, adding a test for HVR2 will reduce the number of coincidental matches.

After both took the test, Bill impatiently waited for the results. The first came in: HVR1 mutations 16183C, 16189C, 16519C; HVR2 mutations 263G, 315.1C. And then the second came in—identical! They were the only two with this exact haplotype in the testing company's database.

The odds of the two samples matching accidentally are obviously miniscule, so based on these results, Bill concluded that Catherine Kelly and Martha Kelly were indeed sisters. He realizes that he can't rule out the possibility of another type of maternal relationship, such as first cousins, but his study of census, estate, and other records points in the same direction as the DNA evidence—and he's delighted to have a new set of third great-grandparents for his family tree!

Identifying the Father by Identifying the Mother

Carey Bracewell has a distant cousin named Brazil whose line stalls pre-1850 because she can't determine the parents of one of her ancestors. The cause of the impasse? Two candidates for the father of this ancestor—a pair of brothers who both lived in the same place at the same time and had daughters of the appropriate age. So which one did she descend from?

Fortunately, it's known that the brothers did not marry sisters, so she realized that if she compared her own mtDNA (she's a direct maternal line descendant) with a known daughter-of-daughters-of-daughters of both brothers, she will be able to assign herself to the right mother—and knowing the right mother will reveal the right father.

Worth the Effort

There are other possible applications, such as determining which of great-great-granddad's kids were born to each of his three wives, but this should be sufficient to give you a feel for what's possible and what's involved. In general, mtDNA testing requires a little more brainstorming effort than Y-DNA. Because

Figure 4-2: Y-chromosome and mtDNA features

Compare and Contrast	Y Chromosome	Mitochondrial DNA
Where found	Males only	Males and females
Size	About 60,000,000 bases; only 23,000,000 sequenced	16,569 bases, all sequenced
Inheritance	From father	From mother
Function	Determines gender	Generates energy
Essential for life	No	Yes
Laboratory technique	Length of STRs	Actual sequence
Cost	Less expensive	More expensive
Polymorphism	Number of repeats	Base substitution
Mutation rate per generation	0.002 per STR (may vary from STR to STR)	0.00001 per base (may vary from site to site)
Sample haplotype format	14–24–12–11–13–13	16192T 16270T
Haplotype diversity	High	High
Longevity of haplotype	Pattern can last hundreds of years	Pattern can last thousands of years
Hereditary disorders	None identified to date	Yes (but not in the HVR)
Identify a person for forensic purposes	Can include some or exclude many, but not unique	Can include some or exclude many, but not unique
Present in hair shaft	No	Yes
Recovery from ancient remains	Very difficult	Possible, depending on conditions

it doesn't travel in tandem with a surname down through the generations, its uses aren't as obvious, and it's harder to organize a project. But, as we hope you've seen from the examples shared in this chapter, it's a powerful tool in its own right and sometimes, it's the *only* tool for the job.

Men and Women/Mars and Venus

Did you used to dread those essay assignments that began with the phrase "Compare and contrast . . ."? It's still a useful device to summarize the many points we've covered in the chapters on Y chromosome and mtDNA testing. Why not dog-ear page 73, and use it as your cheat sheet? We won't tell!

The Broader Picture

The goal of genealogical research is to forge links between specific individuals, generation by generation. But individuals are part of a larger community. DNA tests can place your ancestors in a global framework, even if you have no testing partners.

5

Around the World

Geographic Origins

It's been several thousand generations since the time of Y-Adam and mitochondrial Eve. Can you guess where your ancestors journeyed during that time? Some of them left tracks along the trail that can still be detected today, not literal footprints but distinctive markers on mitochondrial DNA and the Y chromosome.

Chapters 3 and 4 covered the Short Tandem Repeats (STRs) on the Y chromosome and polymorphisms in the Hypervariable Region (HVR) of mtDNA. These markers' mutation rates are convenient for the genealogical time frame and can help sort out close versus distant kin.

But that same mutation rate is too high to preserve signals from our ancient past. If the number of repeats in an STR can decrease in one generation, it can increase in a later generation, leaving it right back where it started (a *back mutation*). If the HVR can mutate at a particular location in one branch of Eve's descendants, it can also mutate at the same spot in another branch (a *parallel mutation*). The picture grows fuzzy if we try to peer back further into the past than a few hundred years for the Y chromosome or a few thousand years for mtDNA. We just can't be sure what the ancestral haplotype was way back then.

If we're being picky, we might wish for some mutations that are so rare that they occur only once at any particular locus (location or position on the chromosome). We'd also need the mutation to be neutral, so that it doesn't affect the offspring. That way, descendants could carry on the new tradition generation after generation, clear down to the present day. After a goodly number of generations, another mutation might occur at a different locus, and still later, a mutation might pop up at yet another locus. Then the chromosome would hold a cumulative record of every change that occurred in that line since Y-Adam or mitochondrial Eve. In the meantime, other descendants of Y-Adam and mitochondrial Eve would be putting their own stamp on the DNA of their lines, with a very low chance of hitting the exact same locus.

One of a Kind

Sound too good to be true? In fact, there is just such a class of mutations, called *Unique Event Polymorphisms*, or *UEPs*. While the word "unique" may be overstating the case somewhat, the mutation rate is so low that any mutation can be treated as a one-time event. Now we have faster markers for the genealogical time frame and slower markers for the archaeological time frame. That's like having your cake and eating it too.

For mtDNA, the slow mutations are in the coding region. Many mutations in the coding region would cause so much damage that offspring couldn't survive. However, there's a small set of mutations that result in just a slight change of meaning, so they can sneak through the fitness filter. It's as if the original version codes for the sentence "The weather will be nice today," while the mutated version substitutes "The weather will be pleasant today," or at most, tweaks the wording to "The weather will be warm today." These silent mutations do not affect survival, so they accumulate gradually over thousands of years.

For the Y chromosome, the slow mutations occur in junk DNA. One common type of UEP is called a *Single Nucleotide Polymorphism* or *SNP* (pronounced "snip"). A SNP occurs when one base (the single nucleotide) is replaced with another base—for example, an A for a G. (By the way, we'll summarize all these mutation-related details in Figure 5-1, so you'll have it for easy referral.) DNA proofreading enzymes are very good at detecting and correcting this kind of mutation, but occasionally one will slip through. There's about one chance in 50,000,000 that a change will occur at any given locus between one generation and the next. Because these mutations are in junk DNA, they can also accumulate over time.

Figure 5-1: Summary of mutation features

Mutations at a Glance		
Mutation rate	Fast	Slow
Historical occurrence	Multiple times	One time (Unique Event)
Y example	Short Tandem Repeat (STR) (e.g., 12 instead of 13)	Single Nucleotide Polymorphism (SNP) (e.g., A instead of G)
mtDNA example	Hypervariable Region (HVR)	Coding region
Time frame	Genealogical	Archaeological
Application	Connections to specific individuals	Connections to broad geographical areas
Results define the ...	Haplotype	Haplogroup
Number of people who share the same results	Small	Large

WHO—WHAT—WHEN—WHERE?

UEPs, although rare, can happen any time. If your father pioneered a new mutation on the Y chromosome, or your mother modified the mtDNA in her egg, you would be the only person in the entire world with that UEP (until and unless you pass it on to your children). If the change occurred in ancient times, chances are you share that UEP with a huge number of distant cousins.

Each mutation first showed up in a specific person who was born at a specific latitude and longitude on a specific day. This was a real person, not an abstraction. If he or she had children of the right sex, who had children, who had children, continuing down to the present day, then he or she became founding father or founding mother of a new clan with a unique DNA marker.

Some of those descendants stayed put, but some migrated in different directions. The UEP is typically found in the highest concentration near the point of origin, for each wanderer must establish a presence, starting from scratch at each new location, while the homebodies keep on piling up more descendants. Global surveys will often show gradients or *clines* in the frequency. This is not an ironclad rule, for if just a few people settle on an island, each individual is automatically a big percent of the population in the beginning (*founder effect*).

"Who" is a specific but unnamed person. "What" is a UEP. "Where" is determined by large-scale population studies with samples of living people from around the world, often from remote, exotic, or even dangerous locales. For a diary from one expedition, check out Spencer Wells' site (popgen.well.ox.ac.uk/eurasia/htdocs/index.html). "When" is trickier, but pooling data from many people can provide some clues. Since the Y chromosome or mtDNA is a cumulative record, we can sometimes determine the order in which mutations occurred. Let's begin with the history of the Y.

Which Came First?

The small data set in Figure 5-2 shows a few markers found on the Y chromosome.

The first column shows which markers tested positive in the samples, and the second lists the locations where the samples were collected. The column for Haplogroup will be explained later. The labels for various UEP markers are not meaningful; you can think of them as part numbers in a catalog. The table already simplifies the task by listing the UEPs in chronological order, but if you wish, you can make the task more challenging by shuffling them. Researchers often rely on computer software to handle data sets with hundreds of UEPs and thousands of samples.

All five samples have the M89 mutation, so they have a common ancestor. They diverge after that; one sample has M172, while the other four have M45. Therefore M89 is older than M172 and M45. M45 splits into two groups, M173 and M242. Some but not all M242 have another mutation, M3, so M3 must have occurred after M242. Some M173 have M17, and others have P25, so those two mutations must have occurred after M173. There's no way to tell from this list whether M17 or P25 came first, just that they are both more recent than M173.

Figure 5-2: Some unique event polymorphisms

UEP Markers	Location	Haplogroup
M89+ M172+	Greece	J
M89+ M45+ M173+ M17+	Norway	R1a
M89+ M45+ M173+ P25+	France	R1b
M89+ M45+ M242+	Mongolia	Q
M89+ M45+ M242+ M3+	Peru	Q3

When this chronological record is combined with geographic hotspots for each pattern, a map of male migration routes emerges. The picture will be refined as more populations are sampled and more UEPs are discovered, but the rough outline is much like that shown in Figure 5-3. The timing of migrations into Europe is not always obvious. The most direct route appears to be M172, but M173 arrived first. Michael Hammer, a geneticist at the University of Arizona, has described Europe as a receiver of intercontinental signals primarily from Asia. Perhaps the old Scottish folksong should be revised to "Oh! Ye'll take the short road, and I'll take the long road, and I'll be in Europe afore ye!"

Figure 5-3: Migration route of some Y-chromosome SNPs

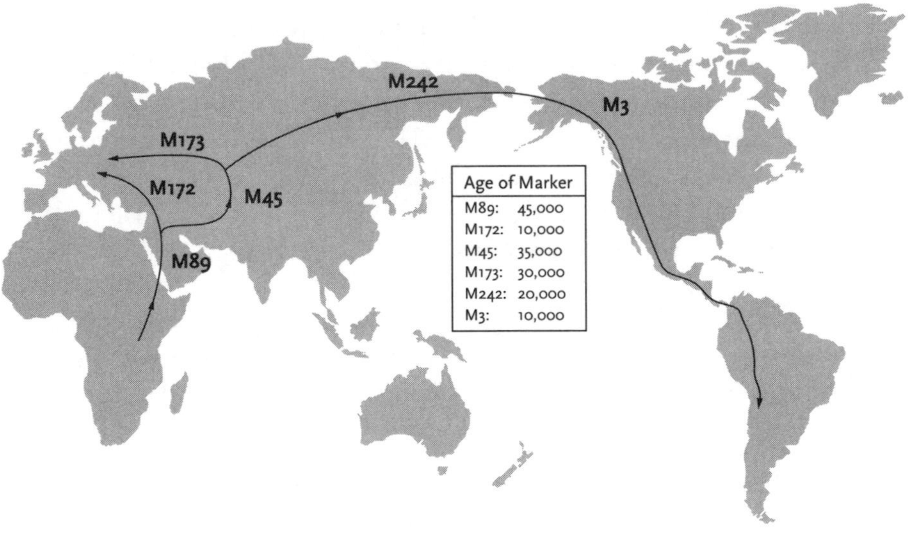

Age of Marker	
M89:	45,000
M172:	10,000
M45:	35,000
M173:	30,000
M242:	20,000
M3:	10,000

DNA "Groupies"

With this background, it's time to introduce a new term, *haplogroup*. Until now, we've employed terms that describe one individual, all with the suffix "type." A phenotype is an observable trait or phenomenon. A genotype is the actual DNA sequence for some region of interest, which can be located on paired or unpaired chromosomes. A haplotype is the complete set of results from multiple sites on a chromosome inherited from one parent (here, the single chromosome is the Y or mtDNA).

A haplo*group* is a cluster of people who share the same UEP, a distinctive marker that all have inherited from a single ancestor. Their ancestry converges on the founding father or founding mother who first sported that mutation. They may or may not have the same haplo*type* (which is based on the fast markers described in Chapters 3 and 4). In fact, if a long time has elapsed since the UEP occurred, there will be many distinct haplotypes within the haplogroup. The age of a UEP can thus be estimated from the level of haplotype diversity, given some assumptions about mutation rates and the length of a generation. For example, the mutation M242 in Figure 5-2 has been dated to about 20,000 years ago. It's found in Asia as well as the Americas, while the M3 mutation is most highly concentrated in the Americas. That puts an upper time limit on the origins of Native Americans—they did not come until some time after the M242 mutation appeared on the scene.

"Hi—I'm a J. What's Your Haplogroup?"

All of mankind has been classified into an outline for the descendants of Y-Adam that shows the branching order of many UEPs. The branches have rather dull names, like the Q3 and R1a found in Figure 5-2. The full *phylogenetic tree,* along with a world map showing the distribution of various haplogroups, can be found in an excellent survey article by Mark A. Jobling and Chris Tyler-Smith, "The Human Y Chromosome: An Evolutionary Marker Comes of Age" (www.le.ac.uk/genetics/maj4/JoblingTS.03.NRG.Review.pdf). Family Tree DNA includes more than just the outline labels with their reports of Y chromosome results. For some examples, see *Sample Y-DNA Haplogroup Descriptions* on page 83.

Is a Best Guess Good Enough?

Although knowledge of your Y haplogroup is not essential for many genealogical projects, where you are looking for recent connections between people, it does flesh out the picture. Haplogroups are generally capable of distinguishing broad geographic categories, European or African or Asian or Native American. This may be a point of idle curiosity (which could turn into a passion as you're stimulated to study more about archaeology and ancient history), or it may provide an important clue if the very continent of ancestry is in question. For example, it may confirm a family legend of Native American ancestry or show that an African American has a European forebear.

Your haplotype numbers seem rather dull compared with the descriptions of faraway places found in Family Tree DNA's *Sample Y-DNA Haplogroup Descriptions* (see page 83), but you may hesitate to spend money on a formal haplogroup test. Fortunately, your haplo*type* numbers can point you in the right direction, even if they cannot give you absolute proof of your haplogroup. Your haplotype also narrows down the range of SNP tests required if you do decide to proceed. How does that help?

The founding father of your haplogroup had a certain haplotype, the starting point for all of his descendants. His descendants have been creating variations on the haplotype, with each marker increasing or decreasing at random. There will be a large number of distinct haplotypes found within the haplogroup, but the central tendency, the core around which the shifts have occurred, sometimes remains visible for a long period. In Figure 5-5, the *modal value* (the single most common value) for the marker DYS389ii is 29 repeats. It's quite possible that it is also the founding father's value.

Sample Y-DNA Haplogroup Descriptions

E3a: Haplogroup E3a is an African lineage. It is currently hypothesized that this haplogroup dispersed south from northern Africa within the past 3,000 years, by the Bantu agricultural expansion. E3a is also the most common lineage among African Americans.

I: The I, I1, and I1a lineages are nearly completely restricted to northwestern Europe. These would most likely have been common within Viking populations. One lineage of this group extends down into central Europe.

J2: The lineage originated in the northern portion of the Fertile Crescent, where it later spread throughout central Asia, the Mediterranean, and south into India. As with other populations with Mediterranean ancestry, this lineage is found with Jewish populations. The Cohen modal lineage is found in Haplogroup J2.

R1a: The R1a lineage is believed to have originated in the Eurasian Steppes north of the Black and Caspian Seas. This lineage is believed to have originated in a population of the Kurgan culture, known for the domestication of the horse (approximately 3000 B.C.E.). These people were also believed to be the first speakers of the Indo-European language group. This lineage is currently found in central and western Asia and India and in Slavic populations of Eastern Europe.

R1b: Haplogroup R1b is the most common haplogroup in European populations. It is believed to have expanded throughout Europe as humans recolonized after the last glacial maximum 10 thousand to 12 thousand years ago. This lineage is also the haplogroup containing the Atlantic modal haplotype.

Q3: Haplogroup Q3 is the only lineage strictly associated with Native American populations. This haplogroup is defined by the presence of the M3 mutation (also known as SY103). This mutation occurred on the Q lineage 8 thousand to 12 thousand years ago as the migration into the Americas was underway. There is some debate about which side of the Bering Strait this mutation occurred on, but it definitely happened in the ancestors of the Native American peoples.

Figure 5-5: Range and frequency of DYS389ii alleles

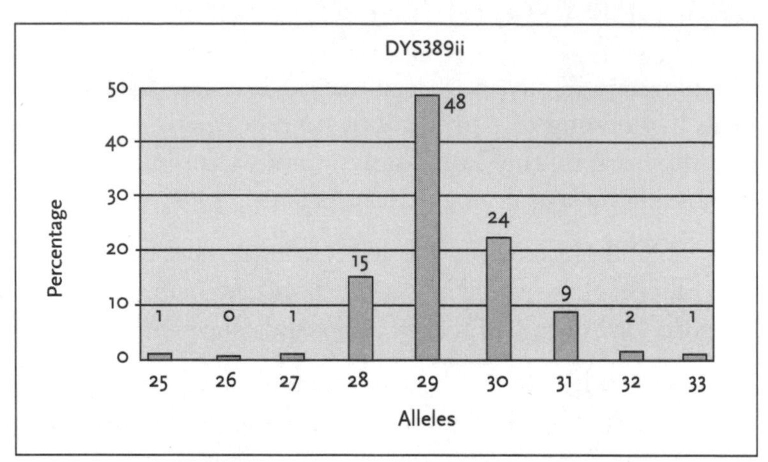

www.ybase.org/statistics.asp

By comparing your haplotype with the most common values found in different haplogroups, you may be able to discern your haplogroup without doing a formal test. The more closely you resemble a modal haplotype, the more confident you can be about your haplogroup. Some genealogical companies include a "best guess" for the haplogroup in their haplotype reports. Other sources of information are the tables compiled by Dennis Garvey (freepages.genealogy.rootsweb.com/~dgarvey/DNA/RelGenMarkers.htm) and the open access databases at www.ybase.org and www.ysearch.org.

In Search of Deep Roots

If some companies provide an estimated haplogroup along with the results of regular Y-DNA testing (i.e., STR testing), why even bother with a formal test for your haplogroup (an SNP test)? Because it's not as obvious for some as others. Many are presented with overwhelming evidence that they are indeed, say, R1b—the most common European haplogroup. They may match dozens of others within a few mutations, virtually all of whom fall into the same haplogroup. (In such situations, some still choose to get the SNP test to be absolutely sure, but we have yet to hear of any surprises.)

But for others, the picture is less clear. When one of us coaxed a male proxy to test on behalf of her Smolenyak line, the closest anyone came to the 12-marker results was four mutations off. Even at this barely linked eight-for-12 level, there were only three matches, and these were listed as I and I1b.

Somewhat paradoxically, there are many rare and unique haplotypes, partially because of the patchiness of the data. While plenty of populations and individuals have been tested, many others still haven't. So if you have no close matches when you get tested, it may be because you truly have a rare haplotype or because your DNA compatriots are presently underrepresented in publicly available databases.

Wishing to gain a little more insight, I went ahead and searched my Smolenyak markers at www.yhrd.org, www.ybase.org, www.ysearch.org, and www.smgf.org. At the first site, I entered nine markers and found no exact matches. Looking for "haplotype neighbors," I found a solitary entry indicating Budapest, Hungary, as the place of origin. One "neighbor" out of more than 20,000 entries. Going to the second site, I entered 25 markers and found no matches. In fact, the closest to my haplotype in the database was a Villanueva who shared 16 of the 25 markers. Delving into statistics for selected markers, I discovered that four of my 25 known markers only appear in 3 to 4 percent of the population. Another two are found in 10 percent or less. I then checked at the third and fourth sites and again found no one even close. All in all, it seemed my haplotype might really be a little on the unusual side, so the odds of my getting an accurate haplogroup estimate anytime soon were not very good. Consequently, I decided to get a SNP haplogroup test.

Several months later, I was classified as haplogroup I, a lineage associated mostly with northwestern and Central Europe and seemingly the second most common European haplogroup. So it appears that I share origins with millions of deep ancestral cousins but come from a branch of the tree that's either sparsely populated or understudied. And since members of our group are thought to have originated in the Middle East about 25,000 years ago and were associated with the Gravettian culture, I now have some pointers to learn more about the early migration of my forebears. It didn't take me long to Google on the word Gravettian and expand my cultural horizons! And as you'll see from the next example, I'm not the only one whose test results have led to the googling of previously unfamiliar words.

Out of Africa

"Hope you can rejoice with me or rejoice for me. I've found an African connection!" So wrote Everett Christmas to his extended family when he received his test results back from African Ancestry (africanancestry.com).

He first learned of the company through a brief blurb in *Parade* magazine in 2003. Researching his family since the 1980s, Everett had stalled out in the early 1800s with his great-great-grandfather. And like many African-Americans, he wished there were some way to learn exactly where in Africa his family had originated. So he requested a kit as soon as he read about the test. Since he had focused much of his research on his strictly paternal line, he decided to order the Y-DNA version of the testing, which African Ancestry calls PatriClan.

The 6 weeks for the testing and reporting of results crawled by, but when the packet arrived, he was so excited that he couldn't bring himself to open it for several more days. African Ancestry had run his sample against their database of almost 10,000 DNA donors from Africa, but even so, three out of every 10 who get tested don't find a match. Would Everett? Finally bracing himself, he was overwhelmed to read, "The Y chromosome DNA sequence that we determined from your sample was an identical match with the Ewondo people in Cameroon."

Ewondo. Cameroon. Everett had never heard of the Ewondo, and although he had traveled to Africa, he had never visited Cameroon. Now both are on his radar. He's learning as much as he can about them, spreading the word to his clan, and planning to visit what he now knows to be his paternal homeland. Even his license plate has been customized to "Ewondo."

African Ancestry uses a Unique Event Polymorphism called a YAP. YAP stands for Y Alu Polymorphism, and it is an insertion of a long, repetitive stretch of junk DNA called Alu. Alus have been inserted at random places all over the genome, but only once on the Y chromosome. It is scored as positive or negative, YAP+ or YAP-. Two of the major haplogroup divisions, D and E, are defined by the YAP+ marker. Haplogroup D is found in Asia, and Haplogroup E is very widespread in Africa. As you might expect, Everett was YAP+ (say YAP as if it were a real word, then add the word "positive").

Once the basic haplogroup is defined with the UEP, African Ancestry checks its database of Short Tandem Repeats for more specific haplotypes found within that large continent. The database has more than 9,000 samples, collected from 60 population groups in West and Central Africa. It does not guarantee an exact match, but it can often find similar haplotypes.

Next on his to-do list? Take the mtDNA version of the test to shed some light on another previously unanswerable question about his maternal ancestry. The same concepts of broad haplogroups and specific haplotypes apply, although different tests are used.

mtDNA Haplogroups: The Daughters of Eve

Bryan Sykes's popular and breezy book *The Seven Daughters of Eve* makes the bold assertion that 95 percent of Europeans can trace their ancestry to just seven women. In technical terms, these women are the founding mothers of haplogroups, prosaically labeled H, J, K, T, U, V, and X. In the book and at the testing company founded by Sykes (oxfordancestors.com), they're more exotically nicknamed Helena, Jasmine, Katrine, Tara, Ursula, Velda, and Xenia. Other haplogroups cover the remainder of the world, as shown on the migration map in Figure 5-6.

Figure 5-6: Migration route of some mtDNA haplogroups

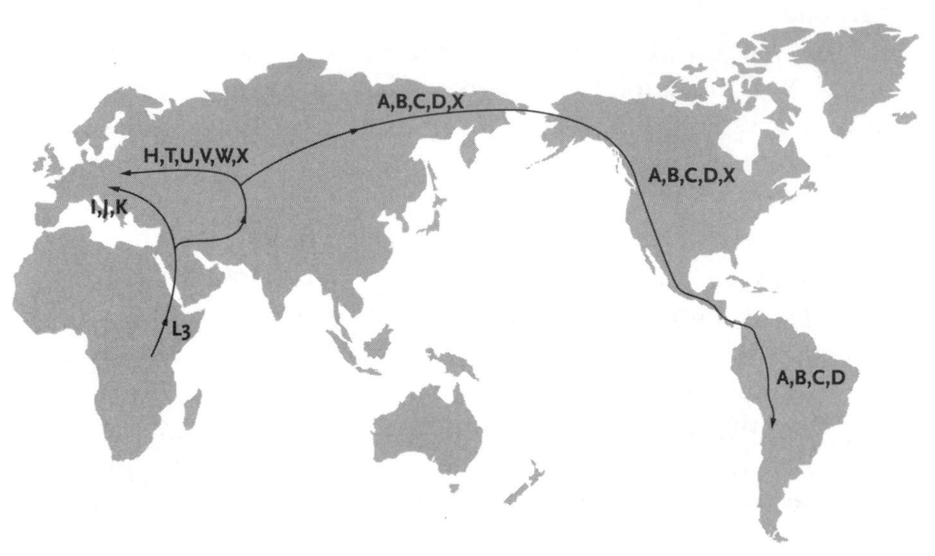

Drawn after a map at MITOMAP to emphasize similarities with Y migration routes. For more details, see mitomap.org/mitomap/WorldMigrations.pdf

Like the Y-chromosome haplogroups, mtDNA haplogroups use an alphanumeric system, starting with a capital letter. The haplogroups are broken down into subgroups with alternating numbers and letters. For example, haplogroup U, probably the oldest of the European haplogroups, has been subdivided as deeply as U5a1a. However, the haplogroup labels were assigned more or less in order of discovery, not organized in a completely hierarchical fashion like the Y-chromosome haplogroup labels. The mtDNA and Y haplogroups do not correlate with each other. For example, the Y haplogroup A is the deepest division, with a root in Africa. MtDNA haplogroups A, B, C, and D were the first to be described, when Douglas Wallace studied variations found in Native Americans.

Testing the Slow Markers on mtDNA

As short as it is, it is still tedious and expensive to sequence all 16,569 bases of the mtDNA molecule. To uncover small differences, early researchers actually borrowed some tools from bacteria, enzymes that chop DNA into smaller pieces. Each enzyme is restricted to sites with a certain DNA sequence, perhaps TCGA. If a person happened to have a mutation in a spot where TCGA usually occurs, then the enzyme would not cut the DNA there. Thus different people will have different fragment lengths, or in the scientific jargon, there will be Restriction Fragment Length Polymorphisms (RFLPs, pronounced riff-lips).

Mutations in the slowly changing coding region, as detected by RFLP testing, constitute the formal definitions for mtDNA haplogroups. Some testing laboratories still offer these RFLP tests (also called enzyme tests), especially for Native American ancestry. However, most laboratories rely on another, newer method: They use HVR mutations as clues for mtDNA haplogroups. For example, when scientists correlated HVR results with RFLP testing for Haplogroup T, they noticed that all the samples had mutations at positions 16126 and 16294. Some had additional changes, but those two mutations were always present. So if a person had mutations at these positions, they would be very likely to belong to Haplogroup T. These two mutations are the *motif* for Haplogroup T. The word motif in the DNA context has the same meaning as a motif in art or music, a recurring theme with variations.

Scientists have correlated RFLP and HVR results for several thousands of samples. It turns out that the "best guess" for mtDNA haplogroups is more reliable than the "best guess" for Y haplogroups. Mutations, even in the HVR, occur less frequently, and an mtDNA haplotype pattern can persist for a longer period of time. The same mtDNA test you might have taken to resolve a particular genealogical dilemma (as we discussed in the previous chapter) will often reveal your mtDNA haplogroup. Since there is a tremendous amount of haplotype diversity, there are some exceptions where a haplotype can't be classified with any confidence.

In fact, looking into one's deep maternal ancestry is the reason most of us take this test. If that's your only motivation, a test for HVR1 only will be sufficient, since that's what the RFLP/HVR databases use to correlate results. Adding HVR2 does increase the haplotype diversity, an important consideration if you want to confirm a connection between two people who have matching results on HVR1.

If you're curious about the mutations that belong to everyone in your haplogroup and the mutations that distinguish you from the crowd, you can look up your haplotype in a table of motifs (www.stats.gla.ac.uk/~vincent/founder2000). If you have all of the mutations found in one row of this table, chances are good that you belong to that haplogroup. Any extra mutations will be your variation on the theme—the more recent additions, which have occurred since the clan mother lived.

From Haplogroup to History

Dr. Ana C. Oquendo Pabón and Rev. Father José Antonio (Tony) Oquendo Pabón, a brother-sister team from Puerto Rico, have compiled a pedigree chart with every slot filled in for six generations. They know the names of all 32 great-great-great-grandparents. Some lines even extend back to the 1600s in Spain and Portugal. But wouldn't you know it; the line that runs out first is the straight maternal line. It ends with Maria Josefa Natal. The Spanish system for surnames combines the paternal and maternal lineages. Thus the surnames for Ana and Tony are derived from their parents, José Antonio OQUENDO Rivera and Rosa María PABÓN Oliveras. María Josefa NATAL, with only one surname, will be harder to trace.

Puerto Rican ancestry can come from several directions dating back to the 1500s: European explorers, African slaves, and the indigenous Taino Indians. Early parish records for marriages and births kept separate books for Blanco (White), Pardo

Figure 5-7: Summary of Eve's seven European daughters

The Seven Daughters of Eve

In his book *The Seven Daughters of Eve*, Bryan Sykes composed imaginative biographies for the founding mothers of the European haplogroups. These are not literally true and factual in every detail, of course, but they serve as a literary device to emphasize the real existence of a specific woman and to place her in a geographical and archaeological framework. The following descriptions are greatly condensed down to the when and where.

H = Helena: Helena's descendants are the most widespread in Europe today. She lived about 20,000 years ago near the border of France and Spain.

U = Ursula: Ursula is the oldest of the clan mothers, one of the first to set foot in Europe, living in Greece about 45,000 years ago.

T = Tara: She lived about 17,000 years ago near the Mediterranean, at a time when much of Europe was covered with ice.

K = Katrine: She lived about 15,000 years ago on the southern slopes of the Italian Alps. One of her descendants is the Iceman.

V = Velda: She lived about 15,000 years ago in the Iberian peninsula. As the glaciers retreated, a few of her descendants pushed northward. Most of the Saami in Finland are descendants of that migration, a founder effect.

J = Jasmine: She lived in the Fertile Crescent during the European Ice Ages, then moved into Europe with the spread of agriculture.

X = Xenia: She lived about 25,000 years ago. Although small in numbers, her descendants have spread into Europe and the Americas.

(Brown), or Esclavo (Slave), but Maria Josefa's children were born after that time period.

Would a mitochondrial DNA test give some pointers to Maria Josefa's origins? Ana submitted her sample, and her report listed thirteen differences from the Cambridge Reference Sequence, far more mutations than anyone on the GENEALOGY-DNA mailing list had ever reported. Ana's brother Tony teased her about belonging to another species. Since he would have the same sequence, that's like the pot calling the kettle black!

Consulting the table of motifs on the Web site, we found that haplogroup L1c has eight of Ana's mutations. That's an African haplogroup, and explains why Ana has so many differences compared with a sequence that came from England. It is simply that Ana is very, very distantly related to the woman who was the source of the Cambridge Reference Sequence.

In many cases like this, haplogroup information will send you scurrying to history books and even old manuscripts. Ana and Tony have found documents where a white ancestor requested permission from the Spanish throne to bring six slaves to Puerto Rico in the 1600s. Could Maria Josefa's ancestor have arrived in a similar fashion? Or did she come more recently from Africa? Mitochondrial DNA cannot answer those questions, but the haplogroup information does focus the area of research.

Are You Descended from a Cherokee Princess?

Without even knowing you, we can answer that question in the negative. The Cherokee Indians never employed the term. But many people do have family legends of Native American ancestry, sometimes vague and sometimes more concrete.

If the ancestry comes from the straight maternal line, an mtDNA test will usually reveal one of the five haplogroups found on the North American continent in Figure 5-6. Haplogroups A, B, C, and D originated in Asia and are seldom found in Europe. As befits its name, Haplogroup X is more mysterious—it can be found in Europe as well as certain areas of North America. This finding led to speculation about an early

European presence in North America, but recent discoveries point to a common Asian origin (near Lake Baikal in Mongolia) for both the Europeans and Native Americans. The two populations have been separated long enough for a new mutation to occur in the Native American side. Trace Genetics has a database of some 4,000 Native American records from many tribes. You can obtain a standard mtDNA report from them, but they will also report if you have matches within particular tribes.

Many people, though, are disappointed when they receive negative results on an mtDNA test for Indian ancestry. Perhaps the family legends have no basis in fact, but sometimes it just may not be clear which line to test. It may be necessary to find cousins descended from different maternal ancestors to fully check the theory. And even if the results are still negative, you have compiled a more complete picture of your different ancestral lines.

The Rest of the Story

Throughout this book, we have been concentrating on the Y chromosome and mtDNA. These two lines are only a tiny fraction of one's total ancestry, but at least they trace a straightforward path. We don't have to guess where the Y chromosome and mtDNA came from: They came down to us as one intact chunk of DNA, in one direct line without any detours, from Y Adam and mitochondrial Eve. But that leaves 3 billion bases worth of DNA to explore on each set of the paired chromosomes (the autosomes), which come one from the father and one from the mother.

Every little snippet of your DNA came from one of your ancestors, but which one? If we could follow it back through time, it would veer back and forth between the paternal and maternal lines in an unpredictable fashion. It could have come from one of your four grandparents, one of your eight great-grandparents, or one of your sixteen great-great-grandparents. Perhaps each little bit could tell us a different part of the story. Genealogy would no longer be limited to the upper and lower lines on the pedigree chart (as seen in Figure 1.1). Every slot in the chart would have an equal chance of contributing a dish to your DNA potluck.

UEPs are not restricted to the Y chromosome and mtDNA. They can occur anywhere in the genome at any time. When we hear the figure that any two humans from any corner of the earth are 99.9 percent alike, that still translates to 3 million differences altogether. Most of these differences are SNPs, a change of a single base, and most of these SNPs occur in junk DNA.

Just as we described for the Y chromosome, SNPs can occur at any time. You and your siblings probably have a handful of SNPs compared with your parents, but the chances are only one out of 50,000,000 that a SNP will occur at any particular locus. It would be astronomically rare for you and your sibling to have a freshly minted SNP at the same location. Your new SNPs are *private*, and so are your sibling's.

But SNPs have been accumulating over the ages. If a SNP occurred before man migrated out of Africa, it could show up anywhere on earth (a *cosmopolitan* SNP), and by the same token, some people who remained in Africa would not inherit the SNP. Cosmopolitan SNPs and private SNPs, the two extremes of ancient and modern, just aren't very useful in tracing ancestral migration paths.

We could wish for SNPs that are found in 100 percent of one population, say Native Americans or Southern Europeans, and never found anywhere else. Here we have finally run out of luck. What's the problem? Consider what happens after a SNP occurs.

If a SNP shows up in one man early in the settlement of the Americas, he now has two versions (alleles) of DNA at that locus, one from the parent who preserved the original value (the same value found in everyone else in the world), and one from the parent who created the SNP. There's only a 50 percent chance that he will pass on the SNP to a child. If a child does inherit it, the same risk applies to the next generation. Most new SNPs will simply disappear after several generations.

Nevertheless, purely by chance, some SNPs do catch on, and after many generations, they appear in a significant portion of the local population, displacing the original allele percentage-wise. In the meantime, some descendants carrying the SNP have struck out for other parts of the world. The net result is not a clear-cut, all or nothing distribution of the SNP in different regions. We have to settle for hints, as shown in Figure 5-8. The marker numbers are from a DNAPrint Genomics test.

These SNPs are "Ancestry Informative Markers," in the terminology used by DNAPrint Genomics (www.ancestrybydna.com). Because the SNPs are in junk DNA, they do not reveal anything about superficial traits sometimes associated with race. The SNPs are selected simply because the frequency in different parts of the world conveys clues about geography. Any one SNP doesn't prove much of anything, but if a whole collection of SNPs points in the same direction, then you can infer that most of your ancestry came from that area.

For example, assume that the base C is the common version and the base T is the rare version for each of the markers in the table. You inherit two copies of each marker, one from your father and one from your mother, but for simplicity's sake, let's just consider the copy inherited from one parent.

Combining the results for all three markers, you might be T-C-T. Then all three markers line up with the alleles typical of Native American ancestry. If you are T-C-C, all three markers line up with European ancestry. If you are C-T-T, all three markers line up with African ancestry. Any one of those patterns *could* be found in the other populations, but they would be less common.

What if you are C-C-C? You have the most common allele in the African population for marker 1136, the most common allele in the Native American population for marker 1141, and the most common allele in the European population for marker 1113. All of

Figure 5-8: Frequency of one allele in different populations for three different SNPs

Marker	African	European	Native American
1136	0.75	0.06	0.02
1141	0.05	0.51	0.75
1113	0.14	0.67	0.05

your ancestry *could* come from just one of those areas, or you could have an *admixture*, with contributions from two or three. By the time you throw the alleles contributed by the other parent into the pot, who knows what ingredients might be in the stew that is you?

Testing a larger number of markers clarifies the picture, but some ambiguities remain. Statistical procedures can arrive at the most likely combination of ancestries to account for your results, but the *confidence interval* is necessarily wide. For example, if your results give you a *Maximum Likelihood Estimate (MLE)* of 35 percent Native American, your actual composition could be 35 percent plus or minus 15 percent. Your siblings may not end up with exactly the same results either. By the law of independent assortment (covered in Chapter 2, Genetic Essentials), every child inherits a different combination of alleles from the parents. The way that siblings differ will be covered in more detail in the next chapter, in the context of relationship testing.

Expect the Unexpected

DNAPrint Genomics is the first company to offer a test based on Ancestry Informative Markers. The test categorizes the results in four broad geographical regions: Sub-Saharan, African, Indo-European, East Asian, and Native American. Perhaps more so than any other DNA testing currently available, results from this BioGeographical Ancestry test can surprise those who opt to try it. For most of us, it just provides a confirmation of what we already knew, but a fair number get back figures that differ somewhat from what they would have predicted. Charles F. Kerchner, Jr., with many documented generations of Pennsylvania German ancestry, took the test out of intellectual curiosity. He received results indicating that he was 79 percent Indo-European and 21 percent East Asian, rather startling since he had expected 100 percent Indo-European. Was the test flawed, or could the East Asian influence be explained by recent historical events, such as invasions of Europe by the Huns, or even by more ancient migrations, such as the path traveled by the M45 marker in Figure 5-3?

In an attempt to shed more light on the subject, Charles developed a Web site (www.dnaprintlog.org), where others could log both their expected and actual outcomes. The following are typical of one cluster of responses borrowed from Charles's intriguing site:

- Doug Mauck also took the BioGeographical Ancestry test out of curiosity. Research on his paternal lines had revealed only European roots, and although both his mother and her mother were adopted, he had assumed that they were likely of European descent. So he thought that the test would simply show 100 percent Indo-European ancestry, but the results came back 90 percent Indo-European and 10 percent Native American, pointing to at least a partially Native American maternal heritage. When he had his mother tested, however, her results were 96 percent Indo-European and 4 percent East Asian, suggesting a need to reinvestigate his father's side of the family.
- Being adopted, Ian Harris only has information about one branch of his family tree, so he decided to take the BioGeographical Ancestry test. Born and raised in Scotland, he figured that a 100 percent Indo-European result was likely, but was startled to find that he tested as 8 percent Native American. He speculates that someone in his ancestral tree was an American or Canadian with some Native American roots.
- Yet another person tested "out of curiosity and a strong sense of my parent's 'hiding' something in my/our ancestral heritage." Anticipating 95 percent Indo-European with a dash of Native American, he instead received results showing a 60/40 split, with a considerably higher percentage of Native American ancestry than imagined.

All three of these involve individuals who discovered evidence of Native American ancestry when they had expected none, but other surprises can occur, either in the categories or percentages that appear. Perhaps the best known is Wayne Joseph, a Californian originally from Louisiana, who was interviewed on *Nightline* for a series on racial issues in America. Joseph had published several opinion articles on the topic when he decided to take the BioGeographical Ancestry test. His results were 57 percent Indo-European, 39 percent Native American, 4 percent East Asian, and zero percent Sub-Saharan African—not especially welcome news to a man who has always considered himself to be African-American.

The most prevalent concern about this test is that many expecting signs of Native American ancestry are presented instead with East Asian results, suggesting that the markers selected to differentiate between these two groups are less than definitive. And given that Native Americans originally migrated to North and South America from Asia, this probably shouldn't be all that astonishing. Still, this is perhaps the most important shortcoming to be addressed through test refinements.

The second most common gripe about this test? Expecting to find a mix of numbers in one's report and turning out to be 100 percent Indo-European, or not finding as pronounced a mèlange as hoped. In some cases, this is reflective of the fact that the test can only reach so far back into an individual's genetic makeup—perhaps a handful of generations. That's because there's only a 50-50 chance that a specific allele will be passed on to the next generation. Those who take the test to confirm Native American ancestry are sometimes disappointed for this reason. But in others, it may well be evidence that the family lore or research is wrong, information that isn't always happily received. In such cases, testing other family members is perhaps the best way to convince oneself of the test's reliability—which will undoubtedly improve with future refinements.

An Embarrassment of Riches

Too many choices? We know we've tossed a lot at you this chapter since we covered several different methods of testing. Each test develops one part of the picture. Sometimes you need a narrow focus, and sometimes you want a broader view. Sometimes more than one type of test will answer the same question. To help you keep it all straight, we've consolidated examples and tests in Figure 5-9.

Figure 5-9: Tests for geographical origins

Typical Objective	Theory Being Tested	Testing Method
Maternal Origins: Learning about your deep maternal ancestry	I wonder which of the "daughters of Eve" I'm descended from.	mtDNA haplogroup
Global Origins: Learning your Indo-European, East Asian, Sub-Saharan, African, and Native American percentages	I'm just curious what my geographic origin profile might be. Are we pure _____ (European, Native American, African, or Asian), or could there be some _____ (same selection) in our family? Dad is of pure Eastern European stock, and I've always wondered if maybe there isn't a little Asian in our heritage due to all those invasions from the East.	BioGeographical Ancestry
Suspected Ethnic Origin: Seeking genetic evidence for family tales	We've always been told that our great-grandmother was Native American. We have some circumstantial evidence, but would like more proof.	mtDNA haplogroup (for straight maternal line of any duration; consider the specialized database at Trace Genetics) or BioGeographical Ancestry (for recent admixture from any ancestral line)
African Origins: Seeking Ethnic/Tribal Origins within Africa	Is our family's male progenitor African or perhaps European due to plantation dynamics? I wonder what tribe my primary maternal/paternal line might have come from in Africa.	Y-chromosome haplogroup (for straight paternal line) or BioGeographical Ancestry (for admixture from any ancestral line) Specialized database at African Ancestry, available for both Y and mtDNA

Critics Corner

While we're on the topic of DNA tests that provide insights into one's geographic origins and "ethnicity," this might be a good time to mention that there are those who are less than enthusiastic about such tests. Regrettably, a few critics have dismissed genetic genealogy as misleading at best (it only represents a small part of an individual's family tree), and harmful at worst (it could reinforce oversimplified or false notions of race and cause identity problems). By contrast, our experience has been that those who involve themselves in genetealogy are well aware of the limitations and more aware than most of the ambiguity of race.

In spite of concerns that we don't grasp the fact that a particular test may only provide insight into one branch of our pedigree, or that another test may only reflect our heritage back a few generations, we are curious to learn what can be learned. Genetealogy is still in its infancy, and those of us who are already practicing it have made our peace with the inevitable learning curve and growing pains associated with being a bit of a pioneer. Focusing only on the limitations is a sure prescription for failure, so why not play with the technology to determine what can be understood now and how to stretch the boundaries of its possible future application?

And as those who have submitted DNA samples, mulled over results, and discussed findings with others can attest, it would be very difficult to dwell in the world of genetealogy and *not* gain an appreciation for how imprecise and even artificial the concept of race is. If anything, we would be the ones pausing to ponder before checking off one box or another (or all of them!) on a standardized form. We were far less likely than most to be surprised that Wayne Joseph's test results conflicted with his self-identification. And we understand better than others how our respective family histories overlap with amazing rapidity and converge as one marches back through time. If anything, genetealogy—like genealogy—has the potential to bring people together because it ultimately helps us understand just how interconnected we all are.

6

Next of Kin

Close Relationships

While many of us are interested in family riddles that go deeper than a couple of generations, we sometimes run into a brick wall very early in our search. Older family members are tight-lipped about something that happened 50 or 60 years ago, and all too often "take their secrets to the grave." There's a reason we all know that phrase!

Maybe there was an adoption or an illegitimacy. Perhaps a first marriage someone would rather forget. Maybe—as Jack Nicholson was stunned to learn in his late 30s—someone's supposed sister was really his mother, and those in the know have lived with the charade so long that they've almost forgotten the reality themselves. These protectors of the past keep their secrets with the best of intentions, but in so doing, prevent others from learning about their own history.

Whatever the situation, these secrets have a way of seeping out—not in full-blown detail, but more as an insinuation. You sense that there's a story there, but you're not quite sure what it is. All you know is there's some doubt when you look at your family tree. And in some such cases, specialized close relationship testing may be able to remove the obstacle so you can proceed with your family history research.

This kind of testing is a little more complicated than the other types we've discussed so far, so let's take a look at how all those forensic tests on *CSI* or paternity tests on *Jerry Springer* actually work. Just a quick aside before we get started—there's some math in this chapter. We promise it's not too challenging, but the jumbles of numbers can look a little intense at times, so it will help if you digest it all step-by-step as we've laid it out.

100 Percent

Who is your closest living relative? The legal definition of "next of kin" varies from state to state, while a dictionary definition may or may not include a spouse. But in a certain sense, the biological and mathematical answer is—you! Every cell in your body shares 100 percent of its DNA with every other cell.

A liver cell looks nothing like a nerve cell under a microscope, and a thyroid cell and a muscle cell have their own special functions, but they all have the same DNA. What makes them so different? Some genes are silent, and others are expressed, yet the DNA is all present and accounted for in every single cell. It has been copied over and over and over again every time a cell divides, starting with one fertilized egg and ending with the trillions of cells in your adult body.

The process continues today: Right at this very moment, while you are quietly reading, you are busy shedding dried-out skin cells to make way for fresh ones, replacing weary old blood cells with energetic young ones, and giving your gastrointestinal tract a shiny new lining. Perhaps a slight change (a mutation) has occurred now and then—nothing is perfect—but you would be hard pressed to locate it among the billions of bases inside the trillions of cells.

Capturing DNA

As you go about your daily life, you are leaving DNA samples everywhere. When you rinse out your mouth after brushing your teeth, your DNA is going down the drain. When you lick an envelope and stick it in the mail, your DNA may travel for hundreds of miles. When you toss an empty can of soda in the garbage, your DNA ends up in the dump. When you take off your baseball cap after a game, your DNA is in the hairs caught in the seams—and even if you carefully removed every single hair, your DNA could be found in the sweat on the hatband.

These trace amounts of DNA might seem impossible to measure—and in fact, they are. Scientists resort to a bit of Hocus-Pocus-Presto-Chango: They mimic the natural activity of a cell when it divides. With the right conditions, the proper enzymes, and ample supplies of simple raw materials, DNA can make perfect copies of itself in a test tube, over and over. Every cycle doubles the amount of DNA. After about 20 rounds of doubling, which might take just a couple of hours, there are more than a million copies of the original molecule. This laboratory trick, called Polymerase Chain Reaction or PCR for short, garnered a Nobel Prize for Kary Mullis in 1993 and revolutionized DNA studies.

But scientists are still not as smart as a single cell. They can't duplicate the entire DNA in all of the chromosomes at once.

DNA from Hair

Mitochondrial DNA can often be recovered from the shaft of a hair. Nuclear DNA, the type of DNA used for the tests described in this chapter, usually requires the root of the hair, the little bulb you see if you pluck a hair. Shed hairs (even hairs found in a hairbrush), or locks of hair cut off as a memento, lack the bulb. Very specialized techniques can sometimes recover small bits of nuclear DNA from these rootless hairs, but best results are obtained from a sample of five to 10 hairs with the bulb intact.

They can only handle short lengths of DNA, a few hundred bases at a time, so they must make a choice about which segments of DNA to work with. For purposes of human identification—distinguishing one person from another—they pick especially variable regions.

Such polymorphic regions are not hard to find, in spite of the fact that we all share about 99.9 percent of our DNA. Just 0.1 percent accounts for all the visible and invisible differences between any two people on the planet. It seems like a tiny difference percentage-wise, but that small fraction results in about 3 million differences. Most of these variations fall into the invisible category: They occur in junk DNA.

DNA Profiles

Fortunately, it's not necessary to check all 3 million differences to uniquely identify a person. A surprisingly small number of tests will suffice. The FBI uses just 13 markers, and even that is more than absolutely necessary. The power of the tests comes from the many possible combinations of results, much like shuffling a deck of cards and dealing out a hand that you've never encountered before.

The FBI markers are sometimes called the CO*DIS* markers, because they are included in the **CO**mbined **D**NA **I**ndex **S**ystem, a database of records with DNA *profiles* from criminal offenders and crime scene evidence. DNA profiles are sometimes called DNA signatures or DNA fingerprints to emphasize their power to identify a person. A DNA profile is simply a series of numbers with no intrinsic meaning or significance, rather like a bar code. The CODIS markers are Short Tandem Repeats (STRs), exactly like the STRs on the Y chromosome, except that they are on the autosomal chromosomes and come in pairs. For every marker, you inherit one copy from your father and one copy from your mother.

You may have a value of 15 repeats for a certain marker, while your brother has a value of 17 repeats. It does not mean that your brother is taller or stronger or smarter (or scored more points) than you. But you do have different alleles (genetic versions). The CODIS markers have such a large number of alleles (seven to 20) that most people have uncommon values, strange as that statement may sound. It's much like surnames: Smith is the single most common name in the United States, yet it only occurs in 1 percent of the population.

The CODIS markers have names like D21S11 (an STR on chromosome 21) and TPOX (named for a nearby gene, thyroid peroxidase). You can think of the names as part numbers in a catalog; it's not necessary to know what the names "mean." You may have caught a brief glimpse of a DNA profile if you watch *CSI* on TV, but here's an example to study at your leisure. Your DNA profile report might look something like Figure 6-1.

For the first one, the D3S1358 marker, you inherited a "14" from one parent and a "15" from the other parent. Your genotype is written as 14, 15. *You cannot tell which parent gave you which value by looking at your own results.* The results are typically listed in numerical order. For the third one, the FGA marker, you inherited two 22 values, so you know both of your parents had at least one 22. The last marker in the table is not an STR. It uses a gene, amelogenin, which is found in different versions on the X and the Y chromosome. It is used to determine the sex of the sample. In this case, the sample has both the X and the Y versions, so it came from a male.

Figure 6-1: DNA profile

Marker	D3S1358	vWA	FGA	D8S1179	D21S11	D18S51	D5S18
Results	14, 15	15, 17	22, 22	12, 14	29, 31	15, 18	10, 13
Marker	D13S317	D7S820	D16S539	TH01	TPOX	CSF1PO	AMEL
Results	11, 12	9, 11	10, 11	8, 9	8, 8	11, 12	X, Y

How Do Forensic Scientists Use These Numbers?

The Royal Canadian Mounted Police (RCMP) has a Web site (www.csfs.ca/pplus/profiler.htm) with a calculator for finding frequencies of individual alleles and the alleles in combination.

If you plug in the values for D3S1358, you'll see that 13.41 percent of the population has a value of 14, and 28.96 percent of the population has a value of 15. That doesn't make you feel very special, does it? But only 7.7 percent (or about one out of 13) of the population will have both values, so we've narrowed the odds a bit. (This calculation is done for you at the RCMP site, but you can figure it out for yourself if you wish. Multiply 0.1341 times 0.2896, then double the answer, since there are two ways of getting a 14 and a 15—14 from your father and 15 from your mother, or vice versa.)

Now check the values for vWA, where 11.89 percent of the population have a value of 15 and 27.74 percent have a value of 17. Again the individual values aren't exactly rare, but the presence of the two markers together occurs in about one out of 15 people. The odds begin to change dramatically when you combine the results on several markers. For the first two markers, only one in 195 people (13 times 15) will match you on all four values. Every time you add another marker, the chances grow slimmer and slimmer that someone else will match you on *all* of them. By the time we have checked just five markers, the chances are less than one in a million that someone else will match you exactly. But there are millions of people in the country, and we want to be absolutely convinced that there is no one else just like you, so we continue to add more and more markers. With these values for the thirteen markers, the chances of encountering your DNA doppelganger are only one in 2,830,000,000,000,000 (that's quadrillions!). Of course, if you have identical siblings, their DNA will be the same as yours, so all bets are off.

Remember: *A DNA profile tells you absolutely nothing about any personal traits.* The numbers have no meaning in and of themselves; it's the combination of numbers that is unique. Think of telephone numbers. How many people in your area code have a number that ends in 7? How many people have a number where the fourth digit is a 3? Lots of them. But by the time you consider the set of numbers as a whole, you are the only person with that combination (or else the telephone company goofed big time).

If your eyes glaze over when you encounter numbers and calculations, consider the clues left behind in a crime scene. How many people have shoe size 8? How many chew Juicy Fruit gum? How many do both? How many of *those* people wear pink lipstick? Every time a new clue is added, the pool of suspects shrinks.

How Are Individual DNA Profiles Used?

DNA profiles can be used to convict the guilty, exonerate the innocent, or identify remains. Searching Google News for the keyword DNA will bring up reports on nearly a daily basis, and there is much controversy about the role of forensic DNA databases in today's society. Some advocate a universal database, while others assert that even prisoners should have the right to refuse DNA tests that could link them to other crimes where they are not suspects.

The Innocence Project, begun in 1992 by Barry Scheck and Peter Neufeld, handles cases where DNA testing of old evidence can yield conclusive proof of innocence. The original convictions in many of these cases were based on erroneous eyewitness testimony or even false confessions. One of the most famous cases is the Central Park jogger, who had no memory of a vicious attack in 1989. Five teenagers confessed to the crime and served time, but DNA evidence later implicated another man. According to the project Web site (www.innocenceproject.org), some 138 prison inmates had been exonerated by the end of 2003.

Many "cold cases" where biological evidence has been stored under good conditions are being re-examined with DNA techniques not available at the time of the crime. Much smaller samples can now be analyzed. The Green River killings near Seattle, with at least 49 victims, occurred in the 1980s, and an early suspect was finally arrested in 2001 after matching his DNA with evidence left behind at the crime scene. DNA is being used to identify some of the victims as well.

DNA profiles are also used in large-scale disasters and mass gravesites, where other means of identification are sometimes impossible. Mass graves from the 1992–1995 war in Bosnia sometimes have hundreds of victims; excavations continue today. The effort to identify remains from the World Trade Center required new techniques especially designed for analyzing small and degraded samples. If there are no reference samples from the victim's own belongings or other sources of DNA, relatives can contribute their DNA for comparison. When Saddam Hussein was captured, DNA testing was immediately performed. The reference samples might have been obtained in a clandestine fashion (perhaps from an ex-mistress), or they might have been DNA profiles obtained from his deceased sons, Uday and Qusay. How can relatives be relevant if everyone is unique? The tools developed for paternity testing (or to be more general, relationship testing) can be applied.

How Is Paternity Testing Done?

Paternity tests often use exactly the same set of markers employed for identity testing. You are unique, and you are the child of two unique people, but your gene pool is more limited than the world at large. Every single one of your alleles came from one parent or the other, and each parent gives you one and only one of his/her own pair of alleles for a particular marker. If you have an allele that neither parent has, then it came from someone else (or, very rarely, there was a mutation).

Suppose you, your mother, and your alleged father are all tested with the 13 CODIS markers, with the results in Figure 6-2.

Figure 6-2: Sample paternity test

Marker	1	2	3	4	5	6	7	8	9	10	11	12	13
"Father"	13,15	16,17	21,22	14,15	29,30	15,15	12,13	12,14	10,11	11,12	9,10	7,8	11,13
Mother	13,14	14,15	19,22	12,13	30,31	16,18	10,12	11,13	9,12	9,10	8,11	8,8	10,12
You	14,15	15,17	22,22	12,14	29,31	15,18	10,13	11,12	9,11	10,11	8,9	8,8	11,12

You can account for every single one of your alleles by looking at your parents. For marker #1, it is clear that your 15 came from your father and your 14 came from your mother. For marker #3, you received a 22 from both parents. These results are completely consistent with a parent/child relationship.

But is that proof? Might there be another man out there who could contribute the alleles that didn't come from your mother? There are many other men who have a 15 for marker #1, and there are many other men who have a 17 for marker #2, and there are many other men who have a 22 for marker #3, and so on down the line. But *all* of those alleles must come from one person. Furthermore, the case is strengthened if some of the alleles are rare in the general population. Paternity labs incorporate frequency data in their calculations for a *paternity index.* Different states have different thresholds for proof of paternity in child-support cases, ranging from a paternity index of 20 (which translates to a 1/20 chance of a false positive, or 95 percent probability of true paternity) to 1,000 (99.9 percent probability). In actual practice, the paternity index often turns out to be much higher than 1,000. For a list of states, see www.dnatestingcentre.com/proof.htm.

Although we have been using the phrase paternity test, the same reasoning applies to maternity testing. Tests for paternity are more common, since the father is more likely to be unknown than the mother (and there are practical issues of child support), but an adopted child who wants confirmation that he has found his birth mother could use a paternity testing laboratory. The actual test is identical.

Adoption Issues

Adopted children exhaust many methods in their search for birth parents. As one part of their strategy, they may sign up for free or low-cost reunion registries where both parties supply some identifying information, such as date and place of birth, in the hopes of finding a match. There are a number of success stories, but matches rely on correct information. What if the information has been lost or garbled or deliberately changed? A registry based on DNA profiles would bypass that problem.

Identigene, a paternity testing laboratory, recently started just such a DNA registry (www.identigene.com/SWIMX/Products/dff_main.asp). Another resource is Linda Hammer, who has made it her life's work to reunite people and devotes much of her effort to the millions whose lives have been touched by adoption. She cites, for example, the fall of Saigon during which 3,000 Vietnamese babies were airlifted out and shipped to Australia, France, Canada, and the U.S. and for which no paper trail may be available. DNA banking offers a new alternative. Those wishing to learn more can visit www.the-seeker.com/dna.htm.

Adoptions and Paternity Testing: Why Bother?

Reunited adoptees and their birth parents say it all the time: I just knew. As soon as we saw each other, we knew. We have the same coloring, the same nose, or the same build. We have the same gestures, the same habits, or the same laugh.

Don Green and Tara Robinson were no different. According to Tara, "As soon as I saw Don, I didn't need a test. We look alike. We chew the same gum. We drink the same coffee." They met when Don, suspecting he might be Tara's father, contacted her. As a teenager, he had an affair with a married woman about twice his age, Tara's mother. Just 4 years later, Tara's mother died from a heart-related condition.

More than 3 decades had passed, when Don, now in his early 50s, had to undergo open-heart surgery. It was then that it dawned on him that Tara could be doubly susceptible to heart problems and should know about her genetic medical history. So he tracked her down, and as soon as they met, it was obvious to both of them that his suspicions were true. But just to be sure, Don's wife ordered kits for paternity testing. Don and Tara's samples were mailed off and, less than a week later, the confirmation arrived. They were indeed father and daughter.

A typical happy ending, right? So was the testing even necessary when the results were so apparent to all? Let's consider another real-life situation. Susan was adopted almost 40 years ago and, like many, reached a point in her life where she wanted to learn more about her birth parents. She was one of the lucky ones. She posted a message on the Internet, and before long, her birth father, John, found it. Their reunion was a warm one and developed into an ongoing family relationship with trips back and forth and holiday visits.

But Susan was still curious about her birth mother, and although John remembered her, he couldn't recall one of the most basic pieces of information: her last name. So Susan's mother remained a mystery until she requested assistance locating her. The intermediary who searched for Susan's mother, Peg, found her less than eager. After a few exchanges, the reason for her reluctance finally emerged in a phone call when Peg revealed that John was not the father. In fact, she claimed not to know John.

Susan and John both reacted with disbelief because it was so apparent to them that they were father and daughter. The trauma of putting a baby up for adoption, the passing of so much time, the turmoil associated with this sudden contact—there were plenty of explanations for Peg's unwelcome pronouncement. Surely DNA testing would put all this confusion behind them. So they ordered the kits, and just like Don and Tara, they had their results in less than a week. John was not Susan's father.

After living as reunited father and daughter for more than 3 years, both were devastated, as was Peg, who had only hoped for the best for her daughter. Fortunately, Susan has a loving family and will likely weather this trying experience with time, but how sad that she should have to. As the intermediary in this case wisely concluded, "It is so important to have DNA testing done, even if you think you are positive of what the outcome will be. There can be no room for error in reuniting families."

Sibling Similarities

In the adoption scenarios we just addressed, we described typical parent/child relationships, but what if an adoptee thinks he has found his birth family, but the alleged parent is deceased? Fortunately, it's possible to DNA test for other close relationships, although it's a little more complicated than parent/child situations. Let's consider the case of siblings.

You share 50 percent of your DNA with your parents or your children. Whichever way you look at it, you give 50 percent of your DNA to each child, or you inherit 50 percent of your DNA from each parent.

Is anyone else as closely related by this mathematical way of thinking? Yes, your full siblings also share 50 percent of your DNA—but this is an *average* number, not an *exact* number. Just why do we say full siblings share 50 percent of their DNA when they have the same parents? "Full" sounds like 100 percent. Pull out a deck of playing cards, a coin to toss, and a pad of paper to record your results as we walk through a statistical experiment.

Suppose we're looking at a genetic marker called Spade. Everyone has two copies of the Spade, one inherited from the father and one inherited from the mother. There are many different kinds of Spades in the general population: King, Queen, Jack, and the cards with one to 10 spots. These alternative forms of Spades are the alleles.

Pick out the Spades from one (the ace) to four, and line them up so that the one and the two are on the left (to represent your father's alleles) and the three and the four are on the right (to represent your mother's alleles).

Now your father is going to create a sperm containing one of his Spade alleles, and your mother is going to pull an egg out of storage from her ovary, which contains one of her Spade alleles. No matter how generous they are, they can't give away

Figure 6-3

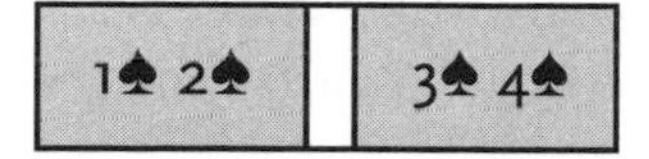

both, because then the baby would end up with four copies—too many!

The choice is completely random, so toss a coin for the sperm. If it comes up heads, write down 1, but if it comes up tails, write down 2. Toss the coin again for the egg, and if it comes up heads, write down 3, but if it comes up tails, write down 4. Now you have the genotype (the two alleles at the Spade marker) of one child.

Repeat this process a dozen times. (We want a large "family" so we can see if there are any patterns.) You'll notice that there are only four possible ways of combining the alleles into genotypes: (1, 3), (2, 3), (1, 4), or (2, 4). Laying out the possible alleles in the column and row headings of a table is one way to ensure you've covered the possibilities.

Figure 6-4: Possible genotypes

		Father's Alleles	
		1	2
Mother's Alleles	**3**	(1, 3)	(2, 3)
	4	(1, 4)	(2, 4)

Each of those four genotypes is equally likely. Did you end up with exactly three of each type? Maybe, but maybe not! Randomly distributed does not mean evenly distributed. Each child was a brand-new blessed event, and your mother and father were not keeping track of the genotypes they had already produced. You *must* toss a coin, not just make up numbers out of your head. Many experiments have shown that people are not good at making up a random number series. For example, it's perfectly possible to see three or four heads in a row, but most people won't include that if they are asked to write down a series of 100 heads/tails that looks random.

Here's one possible series of twelve children (generated by a computer).

Figure 6-5

Child	1	2	3	4	5	6	7	8	9	10	11	12
Spade	1, 4	2, 3	2, 4	1, 3	1, 3	2, 4	1, 4	2, 3	1, 3	2, 3	2, 3	2, 4

Now along comes you, the last child, lucky thirteen. Toss a coin to generate your genotype, just as you did for your 12 older brothers and sisters. Let's say that you ended up with a genotype of 2, 4. Now count how many times you match your siblings. With some siblings you have a 100 percent perfect match, with some siblings you have nothing in common, and with some siblings you share one allele, a 50 percent match.

Figure 6-6

Child	1	2	3	4	5	6	7	8	9	10	11	12
Spade	1, 4	2, 3	2, 4	1, 3	1, 3	2, 4	1, 4	2, 3	1, 3	2, 3	2, 3	2, 4
Match with Child 13	50%	50%	100%	0%	0%	100%	50%	50%	0%	50%	50%	100%

The average of all those numbers is 50 percent. If you had turned out to be (1, 4), the average would be less than 50 percent, or if you had turned out to be (2, 3), the average would be more than 50 percent. But if we looked at a very large sample of many families, the average would be very close to 50 percent.

More Markers

That's just one marker. Suppose we add another genetic marker with four different alleles, the Diamond. The Diamond is on a different chromosome, so it is not linked to the Spade. Your results on the Diamond are independent of your results on the Spade. To keep things straight, let's use the cards numbered five and six for the father, and seven and eight for the mother. If you're ambitious, you can toss the coin again for the 12 genotypes of the Diamond marker. Or you can just look at the series below that we generated with a computer. But don't cheat if you decide to do it yourself—toss that coin to get truly random results.

Figure 6-7

Child	1	2	3	4	5	6	7	8	9	10	11	12
Spade	1, 4	2, 3	2, 4	1, 3	1, 3	2, 4	1, 4	2, 3	1, 3	2, 3	2, 3	2, 4
Diamond	5, 8	6, 7	5, 7	6, 8	5, 8	5, 7	6, 8	5, 7	6, 7	5, 7	6, 7	5, 7

You have the (2, 4) genotype on the Spade, and let's say that you have the (5, 8) genotype on the Diamond. Does your genotype match any of your siblings at *both* markers? Nope, not this time! But another set of random numbers might turn out differently.

Now let's look at how many individual alleles you match:

Figure 6-8

Child	1	2	3	4	5	6	7	8	9	10	11	12
Spade	1, 4	2, 3	2, 4	1, 3	1, 3	2, 4	1, 4	2, 3	1, 3	2, 3	2, 3	2, 4
Diamond	5, 8	6, 7	5, 7	6, 8	5, 8	5, 7	6, 8	5, 7	6, 7	5, 7	6, 7	5, 7
Match	75%	25%	75%	25%	50%	75%	50%	50%	0%	50%	25%	75%

The overall average is 47.9 percent, close to 50 percent, but not exactly 50 percent. That's where the randomness comes in. For a parent/child pair in paternity testing, the number is always *exactly* 50 percent, but for siblings, the number can vary somewhat.

You may have noticed another point when we added the Diamond marker. You lost your exact matches, but you now share at least one allele with 11 of your 12 siblings, everyone except child #9. You have perfect matches on fewer markers, but to compensate, you have partial matches on more markers. You are moving in the direction of sharing 50 percent of your alleles with each of your siblings. If we add another marker, say the Heart, your chances of matching at least one allele improve still further. Child #9 could easily share a Heart allele with you.

How many different combinations are possible with three markers? There are getting to be too many to write out by hand, so we'll make use of a formula. Any one of the four combinations on the first marker can occur with any one of the four combinations on the second marker, then on any one of the four combinations on the third marker. This is sometimes written as 4^3 (four raised to the third power, or four multiplied by itself three times), which is also $4 \times 4 \times 4 = 64$ possible combinations. By the time you get to 13 markers, there are 413 combinations, which works out to be more than 67 million! No wonder siblings can seem so different, even while they share a few traits.

In actual practice, the number of possible combinations is reduced when the parents don't have four distinct alleles. It could happen that one parent is *homozygous* (has two copies of the same allele). This is most likely to occur when an allele is common in the general population. Calculations using allele frequencies come up with a figure of about one in 500,000 siblings matching on all 13 markers.

Figure 6-9

		Father's Alleles	
Possible Genotypes		1	1
Mother's Alleles	3	(1, 3)	(1, 3)
	4	(1, 4)	(1, 4)

We have seen that testing parent/child relationships gives results that are virtually certain. Yet sibling relationships are fuzzier, even though siblings also share 50 percent of their DNA. A

SNPs vs. STRs

The DNAPrint Genomics test for BioGeographical Ancestry, described in the previous chapter, uses a large number of autosomal Single Nucleotide Polymorphisms (SNPs). In principle, relationship testing could use SNPs just as well as STRs. However, SNPs only come in two flavors (bi-allelic polymorphisms, which are present or absent), whereas each STR has multiple alleles. One SNP is not as informative as one STR, since either SNP allele will be shared with a large percentage of the population. It would take more than 100 SNPs to equal 13 STRs in power, but SNPs are easy to measure, and it's possible that they will become more popular for relationship testing. Just as siblings share 50 percent of their STRs on the average, siblings share 50 percent of their SNPs.

Take another look at Figure 6-7 showing the twelve siblings, and imagine that the four of spades and the eight of diamonds are SNPs that point toward Native American ancestry. Some siblings have no Native American pointers, some have one, and some have two. Even though the siblings have the same parents, their apparent percentage of Native American ancestry would end up quite different. Adding more markers will even out the answers, but there is always the potential for two siblings to end up with different results on the BioGeographical Ancestry test.

parent/child combination *must* have something in common on each and every marker (it is *obligatory*), while a sibling pair may or may not have something in common on any given marker.

However, siblings should end up with about 50 percent of their alleles in common, much more than the general population. If some of their alleles are rare, then that also reinforces the conclusion. Paternity testing laboratories will typically couch their reports in terms of probability: "there is a 95 percent chance that these two people are siblings." They may also test more markers than the standard CODIS set, making the cost of the test higher than routine paternity testing. Additional relatives may also be tested: more siblings, or aunts and uncles, or grandparents will add information about the gene pool.

The Romanovs: You Decide

Ready to examine a real-life application of close kin DNA testing in action? For many, their first introduction to the world of DNA testing was hearing about the identification of the Russian royal family. They were killed during the Russian Revolution in 1918, but their burial site was only discovered in 1991. The remains were commingled in a shallow pit. The investigation entailed several stages, with one of the first being the separation of family from non-family. Initially, this was done through the reassembly of skeletal remains and by visually inspecting details such as dental fillings. In this manner, it was determined that the remains probably included those of Czar Nicholas, Czarina Alexandra, three of their children, their physician, and three servants, but further proof was needed. The most publicized aspect of the proof was matching mitochondrial DNA to living kin, including Prince Philip of England. Less well-known is the result of relationship testing, which you're now equipped to evaluate.

Figure 6-11: Autosomal STRs for six sets of remains

MARKER	1	2	3	4	5
Adult Male 1	17, 17	6, 10	5, 7	10, 11	11, 30
Adult Male 2	15, 16	7, 10	7, 7	12, 12	11, 32
Czarina	15, 16	8, 8	3, 5	12, 13	32, 36
Child 1	15, 16	8, 10	5, 7	12, 13	11, 32
Child 2	15, 16	7, 8	5, 7	12, 13	11, 36
Child 3	15, 16	8, 10	3, 7	12, 13	32, 36

Scientists constructed an STR genotype for each of the victims by examining five autosomal STRs from bones from all nine skeletons. By doing so, they hoped to be able to genetically link the alleged Romanovs. As expected, the remains of the physician and three servants were quickly excluded as relatives. The chart on the previous page shows the results from two adult males, the Czarina, and their three alleged children. With your hard-won knowledge about how paternity testing works, can you identify which adult male is the Czar?

Answer: If Adult Male 1 were the father, all the children would have a 17 on Marker #1. (It's an obligatory allele.) There's an immediate contradiction, but continue with the analysis just to show you know what you're doing. For Marker #2, Child 1 and Child 3 are possible, but not Child 2. For Marker #3, the results are consistent with a parent/child relationship, but none of the children have Adult Male 1's alleles on Marker #4. For the last marker, Child 1 and Child 2 are possible, but not Child 3. Adult Male 1 turned out to be the Royal Physician, Dr. Botkin.

How about Adult Male 2? In each case, every STR combination for the possible daughters can be accounted for through the blending of the STRs of Adult Male 2 and Czarina, confirming that the three children were indeed the murdered princesses and that Adult Male 2 was the Czar.

Other Close Relationships

Your next closest relatives are your grandparents, uncles/aunts (if connected by blood, not by marriage), or nieces/nephews, where you share 25 percent of your DNA *on the average.*

If the alleged father cannot be tested, but the mother and both paternal grandparents are available, paternity calculations can be highly suggestive. The grandchild *must* have one of the four possible alleles found in the grandparents at every marker. Relationships between uncle/aunt and niece/nephew do not have this requirement, but at 25 percent, they do share much more of their DNA than two randomly chosen people. Rela-

Figure 6-12: Relationship chart showing percentage of shared DNA

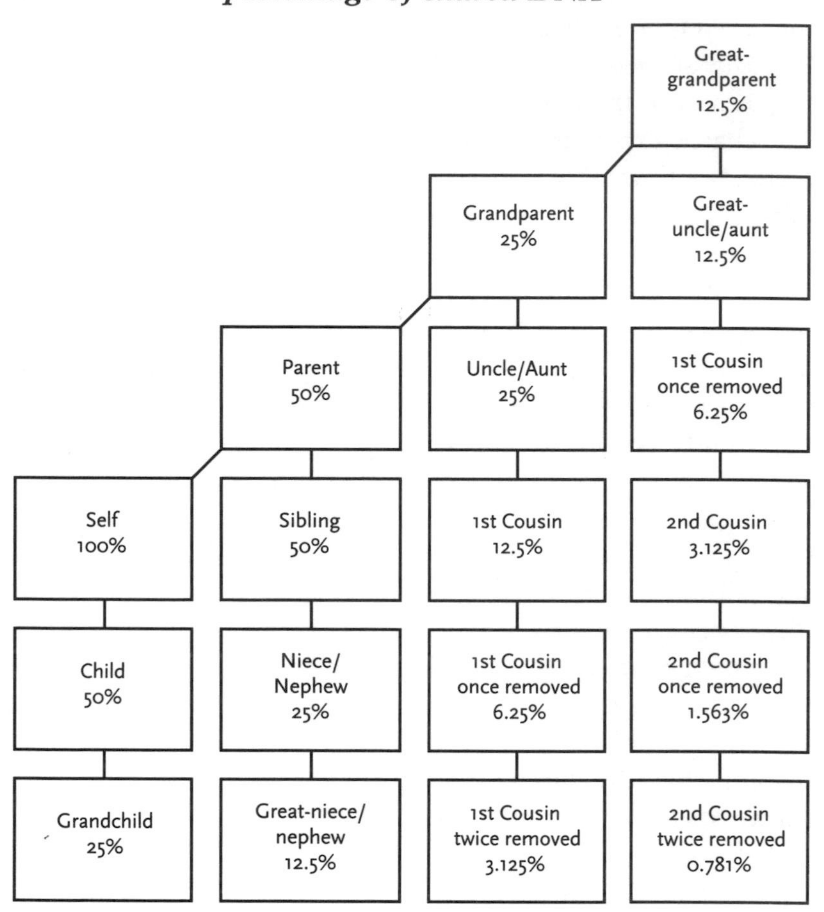

tionship testing can extend perhaps one more step, to first cousins, but that is about the practical limit.

Half-siblings, half-uncles, and other half-relationships reduce the percentages in Figure 6-12 by 50 percent. In 2003, three children of a German hat maker, Brigitte Hesshaimer, came forward with the claim that their father was the famous aviator, Charles Lindbergh. Relationship tests compared them with representatives of the American Lindbergh family, and the results were positive. A single comparison might not be conclusive, but the three children combined would lend more power to the calculations. Since two of the children were male, Y-chromosome tests could also be used to substantiate the paternity index.

Cousin Marriages

While we're on the topic of close relationships, let's take a moment to discuss the ever-popular subject of kissing cousins! Cousin marriages have been commonplace at different times and places in history. Sometimes these have been a deliberate attempt to preserve family wealth or values, while at other times, it has been a simple matter of a limited choice of mates in newly settled territories. Cousin marriages are still the norm in some parts of the world, but they have been declining in the United States as the population grows and migrates and mingles. Currently many states have laws against cousin marriages, ostensibly for medical reasons, but this may be changing as the understanding of genetic risk increases. For a list of states, consult www.ncsl.org/programs/cyf/cousins.htm.

As we have seen, first cousins do share more of their DNA with each other than with a random man on the street, yet the amount is small enough that it is not always possible to prove a connection with relationship testing. This hints at a surprising fact: Cousin marriages are not as harmful as many people suppose.

What are the numbers? The common grandfather of the two cousins may be carrying a harmful gene that does not affect him (a recessive gene, which is not evident unless he has two copies, one inherited from each parent). This is actually a pretty safe bet, since it is estimated that *everyone* carries a handful of deleterious genes. However, his wife, if she is unrelated, probably has a different set of deleterious recessive genes, and their children never get a matched pair that would make the condition evident.

There is a 50 percent chance that he will pass this harmful gene on to his first son and a 50 percent chance that he will pass it on to his second son.

Now those two sons have children of their own. Number 1 son has a boy, with a 50 percent chance of getting that gene. Number 2 son has a girl, with a 50 percent chance of getting that gene. By this time, the chance that both of these cousins have a copy of the gene is 0.5^4 ($0.5 \times 0.5 \times 0.5 \times 0.5$, or 6.25 percent). That is, all four people in the chain must have the gene in order for it to show up in the next generation. Now the boy and girl grow up, decide to marry, and have children of their own. They may pass on that gene to these children, but only one out of four will inherit two copies. One-quarter of 6.25 percent is about a 1.6 percent chance of any specific problem occurring. Yet there is a common misconception that first cousin marriages will inevitably cause problems.

Did You Inherit *Any* Genes from Your Famous Ancestor?

Some of us have kissing cousins in our family trees, while others have someone famous (and some of us are lucky enough to have both!). Can you claim bragging rights if you have a famous ancestor? Sure, why not? But is there a genetic basis to your claim? You almost certainly have inherited some of your genes or genetic markers from your great-grandparents, but a glance at the relationship chart (Figure 6-5) shows there's only one chance in eight that you inherited a specific one. If your great-grandfather was fleet of foot, or your great-grandmother was sweet of voice (traits that may have a genetic component), there's no guarantee that those genes survived the journey down to you.

The farther back you go, the less chance you have of inheriting any particular trait. By the time you get back to 10 generations (your great-great-great-great-great-great-great-great-grandfather), it's quite possible that you have inherited nothing from him. The average amount would be less than one part in a thousand, but big chunks of the chromosomes go by the wayside every time a sperm or egg selects which half of the genetic material to pass on to the next generation. As the amount of DNA from one ancestor gets smaller and smaller, the chances increase of losing it all in one fell swoop.

And yet, every single one of your genes has come down to you in an unbroken line for thousands of generations. That is your legacy, preserved in the face of great perils by some unnamed and unknowable fraction of your ancestors. As Richard Dawkins phrased it in his book *River Out of Eden*, "All organisms that have ever lived ... can look back at their ancestors and make the following proud claim: Not a single one of our ancestors died in infancy."

Close Kin Mysteries

So what if one of those ancestors who survived infancy left you with your own personal history mystery? It just may be that close kin testing can help you over this hurdle. When we say "close kin" here, we are using this as an umbrella term to refer not only to paternity tests, but also to ones for siblingship (full or half), grandparentage, and avuncular (aunt or uncle) or first cousin relationships. The key question for our purposes is how they might be used in a genealogical sense. Here are a couple of actual situations that have confronted researchers and how they could be addressed through close kin testing:

- Brett always knew that his father's mother had been adopted, but by the time he became interested in learning about his roots, she had already passed away. An avid convert to genealogy, he managed to locate what he believed to be his grandmother's birth family in Colorado, but all her alleged siblings were also deceased. However, several of them had children, and Brett made contact. Their family tales were tantalizingly similar, with rumors of an aunt who had been given up for adoption, but all those who knew the truth were gone. How could they be sure? In this case, first cousin testing was an option. If the families had truly found each other, Brett's mother and one of her supposed Colorado cousins could be tested to see if their autosomal DNA profiles substantiated or disproved the family lore and research.

- Larry never bothered to get a copy of his own birth certificate until he retired. When he did, he was astonished to see that his mother's maiden name was the same as that of the foster family who had raised him. They had never spoken of his mother or any family connection, and now there was no one to ask. He did some preliminary research, learned of an upcoming reunion, and decided to attend. When he did, he compared notes with others and learned that his foster father had had a niece, Elizabeth, who fit most of the profile that he had constructed of his birth mother. Could she have been his mother? As it happened, another attendee at the reunion, Jane, was Elizabeth's granddaughter, and she and Larry felt an instant kinship. Unfortunately, both Elizabeth and her daughter, Jane's mother, had died. If their suspicions were correct, Larry and Jane were uncle and niece, so avuncular testing was an option. And if you're particularly nimble with mentally conjuring up family trees, you may have already realized that mtDNA testing was also an alternative since Larry and Jane would have shared a common maternal line, if they were correct about their relationship.

As in the last example, there are some scenarios that can be addressed through more than one type of testing. The Abuelas of the Plaza de Mayo who seek their missing grandchildren—adopted out after being born to mothers who were kidnapped and killed in Argentina's "Dirty War" from 1976 to 1983—are assisted by researchers who use a combination of several types of testing, including Y-chromosome STR analysis (as covered in Chapter 3), mtDNA analysis (as discussed in Chapter 4), and autosomal STR analysis for close kin testing. Not all types of testing are applicable for all situations, but close kin testing is viable in virtually all. By comparing frequencies of selected autosomal alleles of the alleged grandchild with those of his or her suspected grandparents, aunts, uncles, cousins, and so forth, it is possible to calculate the probabilities of relatedness.

While most of us cannot hope to tap into the brainpower and resources of Mary-Claire King, who spearheaded much of the testing used to identify these missing grandchildren, we can do so on a lesser scale. If you have a close kin mystery in your own family tree—and are willing to pay more than most (expect prices in the $450 to $650 range)—you may be able to resolve that question that's hounded you for years.

Selecting a Laboratory for Paternity Testing

There are literally thousands of paternity testing labs with Web sites listed at Google. The yellow pages in your phone book will probably have some listings too, but you do not need to limit your search to local companies. The laboratory can mail a DNA collection kit, usually with a swab to rub inside your cheek, and you can mail the sample back.

There may be two options when you order a test: a legal test, and a "curiosity" test done to satisfy yourself. The test itself is exactly the same, but the legal version requires a third party to collect the sample, along with proof of identity. The chain of custody must be preserved if the results are to be admissible in a court of law; otherwise, a defendant might argue that samples had been switched somewhere along the line.

The American Association of Blood Banks (AABB) has an accreditation program for paternity testing labs. Not so long ago, blood types were a crude method used to exclude parentage, and the AABB has continued to monitor testing methods. Only about forty laboratories are accredited throughout the country. For a current listing, check the AABB Web site at www.aabb.org/about_the_aabb/stds_and_accred/aboutptlabs.htm.

Some smaller companies will send the samples to an accredited laboratory for processing. This is not necessarily a drawback; the smaller companies may have experts on the staff who are skilled at interpreting the numbers and will take the time to explain the results. Other small laboratories may perform the work themselves. Since paternity testing is fairly routine, and commercial kits standardize the process, lack of accreditation is not a fatal flaw for curiosity tests. However, tests for legal purposes might be challenged if they were not performed by an accredited laboratory.

One point of comparison to check when you are evaluating companies is the number of markers routinely tested. Some companies will test a few markers at a time, just until the paternity index reaches a minimum number, typically 100. Other companies will routinely test the 13 CODIS markers or even more.

A Little More Math

This chapter is the most mathematical one you've encountered so far. The trend continues in Chapter 10, Interpreting and Sharing Results. But don't despair—there are some nuggets tucked away where the only math required is counting!

Part 3

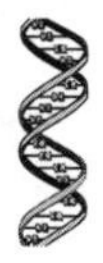

How to Do It Yourself

7

Joining or Running a Project

This all sounds pretty intriguing, doesn't it? A chance to take a peek into your past by simply providing a DNA sample. And getting involved is easier than you might have thought.

The Joys of Joining

Back in the dark ages of genetic DNA testing—way back in 2001 or 2002—the answer would have been to launch your own study. And depending on your circumstances, that might still be appropriate. But for more and more of us, simply joining an existing project is becoming a viable option.

As we mentioned in the opening chapter, genetealogy has reached the tipping point, the phrase Malcolm Gladwell coined for when a concept "crosses a threshold, tips, and spreads like wildfire." And that's good news—especially for newcomers. With the proliferation of surname projects over the past few years, there's a reasonable chance that you'll be able to find one or more projects already in full swing that conveniently focus on names that appear on some branch or other of your family tree. In such cases, rather than start from scratch, you can simply jump into the Johnson, Kincaid, or [Insert your own name here!] project.

Finding a Surname Project

There are well over a thousand surname projects under way. And if you sport an especially common name, such as the U.S.'s current top five—Smith, Johnson, Williams, Jones, and Brown—you may well have several projects from which to choose. In fact, new projects have begun to sprout so quickly that one of us has been able to join projects for two branches of her family tree—Shields and Reynolds—in just the past month. And don't think that just because you have an unusual name, there's no hope. Surnames already under study include Blueglass, Janita, Mockensturm, Quiden, Tuxhorn, and Zuraff.

So how do you find out if there's a project for you to jump into? You have several options, most of which involve the Internet, so if you don't have a computer at home, you'll probably want to visit the local library. If you happen to be an avid genealogist, chances are that you will soon stumble across mentions of projects involving names of interest in the magazines and newsletters you read. Each issue of the National Genealogical Society's *NewsMagazine,* for instance, includes announcements of assorted projects seeking participants. If your name is Hill, Shugart, Sisson, Taliaferro, or Stiddem, you could have recently tripped across a project this way. Publications of societies and family associations—and increasingly, even "regular" periodicals ranging from *Newsweek* magazine and *The Wall Street Journal* to your local paper—contain more and more articles about the launch or progress of DNA studies. Wells, Howery, Glennon, Johnson, Rice, Clough, and Surdival are just a few of the names you could have discovered by flipping the pages of a magazine or newspaper. If you haven't already seen such announcements or articles, you can expect to find them in your mailbox in the not-too-distant future.

Even if you subscribe to several family history publications, you'll want to go surfing on the Internet as well. Many surname projects have their own Web sites, so a "surname DNA" query (replace "surname" with the name of interest) at your favorite search engine will frequently pop up exactly what you're looking for in the top listing or two. For example, if you search on "Blair DNA" at www.google.com, you'll be presented with several links to a Web site with all the details you need. Even if you don't find immediate links to such a site, you'll usually discover a few messages about the project. Click on these, and then simply e-mail the person who made the posting or the project manager, who will often be identified within the message.

You might also want to peruse the project listings such as those provided at www.dnalist.net. Public access databases that are searchable by surname and provide contact information for submitters (www.ysearch.org and www.ybase.org) offer another alternative. If you still find no obvious links to an existing project, you'll want to go to the Web sites of the testing companies, and use their search functionality. For instance, www.familytreedna.com, www.relativegenetics.com, and www.dnaheritage.com all have a search-by-surname option on their home pages. By entering a name here, you will learn how many individuals of that name have already been tested by that company. This is especially helpful for finding small, private studies or recently initiated projects that may not have been publicized or don't have their own Web sites yet. Be sure to try variations of your name (e.g., Strickland, Stricklin, Strickling) or entering just the first few letters to get a more complete list of names that may be close to your own. Because the spelling of our names has often been slightly tweaked over the centuries, it may be worth joining a project based on a name similar to your own, even if it's a letter or two off from what you consider to be the "correct" spelling. Just because you and some fellow in New Zealand spell your name a little differently doesn't mean you didn't have the same great-great-great-great-great-grandfather!

Joining a Project

Let's say you've poked around a bit and found a project you'd like to join. If it's a well-established one with a Web site, all the information you'll need will usually be on the site. You'll be able to learn about the scope of the project—all Taylors or just the ones from certain states?—have a look at results to date and how they're shared, see if the administrator has any special requirements, and of course, join the study. (Incidentally, this might be a good time to mention that project administrators use a variety of terms for themselves, so we use the words administrator, coordinator, and manager interchangeably.)

Almost all such Web sites will include a link for ordering your kit directly from the testing company. This is especially convenient because much of the coordination has already been handled for you. By using the link provided, you're automatically included in that particular study, and the administrator will be notified of your participation, even if you don't contact him directly. Another possible advantage of joining an existing group is a price break! Some companies offer those testing through a project prices that are about 25 to 40 percent less than those testing on their own.

If you've decided to participate, it's simply a matter of following the link and typing in your name, contact, and payment information. Most companies accept online credit card payment, and some will give you the option of being invoiced, so you can wait until returning the kit to part with your money.

Special Requirements

Back in the early days of genetic genealogy, taking and paying for a test was usually all that was required to join a project, and in many cases, that's still true. Every additional test moves a study forward so project managers are usually more than happy to welcome new participants. But as with any practice that becomes more established and verges on entering the mainstream, the world of genetealogy has developed some formalities. Some administrators now have additional requirements for joining. And while you might grumble at the extra paperwork and think it smacks of bureaucracy, it's actually beneficial, even if it takes a little more time.

The two most typical extra requirements make a lot of sense. One is information about your earliest known ancestor and how you (or the person being tested, if it's not you) connect to this person generation by generation. This is frequently provided in the form of a pedigree or descendancy chart, although many project coordinators will take it in any form it's offered. Requesting this information allows the administrator to make sure that would-be participants understand all the necessary nuances—a woman expecting to represent her line in a surname project would quickly be advised to find an appropriate male relative to take the test instead.

It also provides more value to everyone in the study. Think about it. If your test matches someone else's, what's the first thing you'll want to do? More than likely, you'll want to find out as much about this other person and his ancestors as possible. At a minimum, you'll have found a new cluster of cousins, and in an ideal situation, you'll be fortunate enough to match someone who has their family tree traced back a century or two further than you have, so you can benefit from his research!

Most administrators who gather such information from their project's participants also post it on their Web sites. This means that you won't have to wait for your newly discovered cousin to return from his vacation to exchange e-mails and explore possible connections. It will already be available for viewing on the Internet, which leads us to the second frequent requirement—the consent form.

When you take a test, you'll have the option of signing a brief consent form for the testing company. Not signing it almost defeats the purpose of testing, as this release allows the company to notify you of any matches in their database. It's innocuous, and without it, you'll severely restrict the potential to gain insight into your roots, so it's best to simply sign this one. The secondary consent form requested by some administrators is intended to give them permission to post some of your family data on the Internet, but don't let this alarm you. If you read it closely (and most consents are a page or less), you'll discover that the administrators usually restrict themselves in terms of what they're allowed to share, meaning that your data will only appear disguised under a code of some sort or perhaps the name of your earliest known ancestor. In this way, your privacy is protected.

Additional clauses are generally designed to avoid misunderstandings and may cover such issues as these:

- participants being appropriate for the study (e.g., having the right surname or line of descent)
- acknowledgment that there may or may not be any matches, especially in the project's early stages (to avoid possible disappointment when the results are returned)
- who pays
- timely response (so the coordinator doesn't have to chase down stray kits).

Many now also include an indemnification clause to protect the project manager from any claims of harm. To date, we know of no such claims ever having been made, but in our litigious society, it's not unreasonable for administrators to take this measure.

When There's No Web Site

What if you've discovered a surname project, but can't find additional details? Perhaps you've learned of it by getting a hit at one of the testing companies' sites, but there's no indication of a dedicated Web site. In this case, you may wish to do some surfing on relevant surname mailing lists and message boards to see if there's been any mention of a DNA project for this name.

If so, send an e-mail to those who discussed the topic asking how to learn more.

Family Tree DNA's Web site includes a convenient feature that permits you to communicate with the administrator of an existing project of interest to you or to join the study by simply ordering a test. Other companies will presumably offer this feature in the near future, but in the interim, you might consider contacting the company itself to see if they would be willing to play middleman and put you in touch with other clients of the same name. Alternatively, you can opt to order a kit on your own. While you will lose some of the advantages of participating in a formal project, you will still be informed of others who match you (exactly or closely) in the company's database, whether they share your name or not.

Launching Your Own Project

What if you've searched everywhere but can't find any indication of anyone of your name having been tested? Or perhaps you want to study the Smiths of Oklahoma, and there are only projects for New England and Southern Smiths. Or you want to venture into less-charted territory and study anyone hailing from Roseto, Italy, rather than just those with the name Marino. Or you're more interested in mtDNA than Y-chromosome testing. Or you have a brick wall that can't be cracked through a standard surname project. Or maybe you're just a rugged individualist who wants to run the show. For whatever reason, you've decided to launch your own project. Now what?

We were in this same situation a few years ago when we both launched our first genetic genealogy investigations. And we learned by trial and error—as people in such circumstances often do. So we'd like to think that we have more than a few insights to share about running a project. But realizing that the experience of two people couldn't possibly encompass everything this strange, new world of DNA roots-testing can toss at a project manager, we decided to conduct a survey of administrators to gather their advice, look for patterns, and learn about innovative tactics they might have developed.

We deliberately sought out a cross section of managers, being careful to include projects that ranged from one to more than 150 participants, were clients of various testing companies and institutions, had been initiated anytime between the pioneer days of 1998 to the previous week, were local to international in scope, and had varying objectives. Not surprisingly, most of them were surname-focused, but we succeeded in finding a few that weren't.

In all, we contacted 100 project managers, and it speaks of their dedication that 50 (collectively running 60 studies) were willing to take the time to answer our rather long questionnaire. Blending the wisdom from this road-tested group, we arrived at the four-step process for effective project management that will be the focus for the balance of this chapter. If you are new to the world of DNA testing, we are confident that it will considerably shorten your learning curve. And even if you're an old pro (as much as anyone can be in this arena), we hope that you will benefit from the insights and experiences of your peers. We certainly did!

Figure 7-1: The DNA project management process

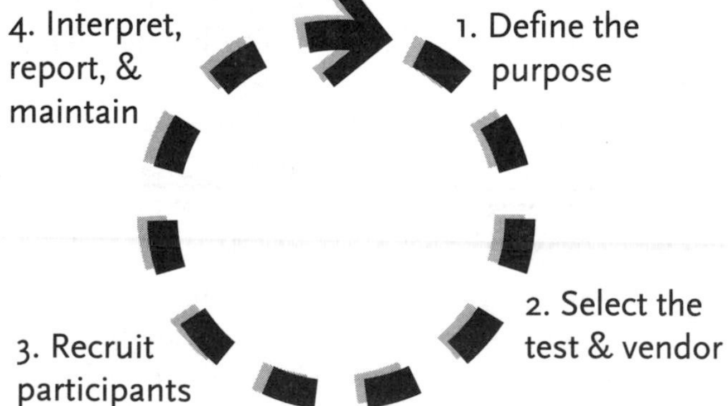

The Project Management Process

As Figure 7-1 shows, the process includes four steps that are generally cyclical in nature:

1. Define the purpose
2. Select the test and vendor
3. Recruit participants
4. Interpret, report, and maintain

We've chosen to portray the process this way because some may have a single genealogical dilemma to tackle and succeed in finding an answer after one trip around this cycle, but many find that the first round just whets the appetite for more! It is not at all unusual for first-timers to press on past the last step back to the first in the hope of obtaining more insight or trying to solve another genealogical riddle.

And while more established projects will frequently find themselves in a mini-cycle rotating repeatedly through the third and fourth recruiting and reporting steps, even they can be sent back to step 1 by an unexpected finding or to step 2 by the introduction of an upgraded test. Assuming you plan on managing a project of more than 10 or so participants, though, you can expect to spend the majority of your time on the last two steps. For that reason, while we will cover the entire process in this chapter, we'll take a closer look at the recruiting aspect in Chapters 8 and 9, and interpretation and reporting techniques in Chapter 10.

1. Define the Purpose

As we discussed at length in earlier chapters about the various types of tests available, it's critical to ask the "Why?" question up front. Why are you testing? What are you hoping to learn? If you haven't already done so, you can be sure you'll be forced to crystallize your objective once you start recruiting participants who will inevitably quiz you about what you're trying to accomplish with your DNA project.

To help focus your thinking, we suggest you revisit the *Typical Y-Chromosome Testing Objectives* chart on page 42 or the *Tests for Geographical Origins* chart on page 99 if you're contemplating something other than a standard surname project. Use these tables like a menu. Peruse them, and pick what's best for you—and just as with menus, you may find yourself drawn to more than one item. That's fine. Some projects have more than one objective. But knowing what you're aiming for and being able to articulate it is a key first step in running a DNA project.

2. Select the Test and Vendor

Once you've defined the purpose of your project, you're ready to move on to the second step of selecting the appropriate test and vendor. While you might expect these to be two, distinct steps, we've combined them because they are inextricably intertwined. Your objective all but dictates which test you'll use, and since certain tests are only available from certain vendors, your test selection will largely drive your choice of vendor.

To help you see at a glance which companies offer which tests, we've provided a vendor list in Appendix B. This list will point you in the right direction, but we caution you that this area is such a fast-moving target—with new companies entering the market and constant shifting in their offerings—that the situation may well have changed by the time you view the list!

You'll notice that a couple of tests are only available from one vendor. For those seeking to learn about their African heritage, for instance, your decision is already made. By contrast, if you're planning on doing strictly Y chromosome or mtDNA testing, you have multiple options. If you anticipate using several types of tests over time, you'll want to factor this into your initial decision because migrating a project from one company to another can involve a great deal of effort and cost.

It's also important to know that tests other than the standard Y-line and mtDNA are more fluid in terms of their availability, so you might want to do a little Internet homework before ordering any specialized DNA tests. You may discover that one that was available 6 months ago is no longer an option, or that one that didn't exist 6 months ago now is, or that the companies selling a particular test have changed.

Another test-related factor is the relative sophistication of the products provided by these companies. For example, some companies only offer one Y-chromosome test for a set number of markers, which may be more or fewer markers than its competitors. As we discussed in Chapter 3, Male Bonding, more markers translate into greater cost but also more meaningful results. You may opt for the company selling the most advanced tests for one that allows you to stair-step your expenditures by starting with low resolution tests and later upgrading, or for one that lets you pick exactly which markers you'd like tested. If you want state-of-the-art testing or more flexibility, you'll want to be sure the company you're zeroing in on can provide it for you.

Beyond the Tests

Should you be among the lucky majority having several vendors to choose from, what else might you want to consider before making your final selection? Here are the factors most frequently cited in our survey of experienced project administrators:

- **Turnaround time**—How long does the company typically take to process the test that interests you?
- **Responsiveness**—When you send an e-mail or try to call with questions (or one of your participants wants to do the same), do you get a response? How quickly? Does the company continue to work with you until the matter is satisfactorily resolved or does it seem to be in a hurry to shrug you off?

- **Reporting**—What will the company give you besides the standard report with its plethora of numbers? Does it provide any additional analysis online or in hard copy? Does it provide any tools such as Web site templates that make it easier to share results with others?

- **Database Access**—Does the company provide a matchmaking service so you'll be sure to learn of any matches—in or out of your own project—for your participants? How large is its database, and how fast is it growing? Will your access be indefinite, or is there a time limit to it? Can you search the database by surname?

- **Management Tools**—As the project manager, will you have to play middleman in every transaction, or does the company allow participants to order and pay for kits directly? Will you be notified of kit receipts, processing, and results? Will you have access to results for all your participants, and are there any study-wide reports available?

- **Cost and Payment Options**—What is the basic cost for a test? Are there any discounts? If so, how many must participate? Can you pay by check or credit card? Can you order online? Can your participants place their own orders? Can one person order the kit, but have it sent to someone else at another address?

- **Sample Retention**—How long will the company retain samples submitted? When they offer new or upgraded tests, can that same sample be used, or will it be necessary to submit another? (Or would you prefer to work with a company that destroys samples after testing?)

- **Special Services**—Does the company offer any additional services, such as the option to preorder a batch of kits for a reunion? Do they offer forensics services to obtain DNA samples from hair, envelopes, and so forth? If not, can they at least test samples obtained in this manner by other labs? What else differentiates this company from its competitors?

- **Pushing the Envelope**—What is the company's reputation for being a leader or follower? Do they periodically offer new or upgraded tests? Are they initiators, or do they only push forward when more or less forced by competition?

The best way to get answers to these questions is to ask those who were in your shoes not too long ago—that is, experienced project managers. You can easily do this by posting a message to the GENEALOGY-DNA Mailing List (http://lists.rootsweb.com/index/other/DNA/GENEALOGY-DNA.html), a favorite haunt of project administrators. Alternatively, you can contact coordinators directly by e-mail to ask for their advice. The same techniques outlined in the *Finding a Surname Project* section earlier in this chapter will help you locate these people with minimal effort.

3. Recruit Participants

At this point, you have an objective in mind and have made the necessary decisions about tests and vendors. Now it's time to recruit participants! Successful recruiting requires two key ingredients: the skills to find appropriate people and the ability to deal with them. Both aptitudes are critical because if you have no participants, you have no project—but they have surprisingly little to do with each other. A study coordinator could, for instance, be an ace at locating appropriate participants and terrible at persuading them to join a project, or vice versa (which is why some projects wisely assign these responsibilities to different people).

For this reason, we've devoted a chapter to each aspect. In the next chapter, we'll discuss two fundamental strategies for finding participants—proactively seeking them out or making it easier for them to find you. Once you've found a likely testing candidate, it's tempting to immediately send an e-mail or make a phone call, but requesting a sample of someone's DNA is not a routine activity! A little preparation can greatly improve your chances of success, so in Chapter 9, we'll discuss potential complications, such as common myths, privacy concerns, and money matters.

4. Interpret, Report, and Maintain

On the surface, it seems that your work is done once you've had people tested, but in most cases, you have a little more to do. For many, reporting and interpretation are not the most thrilling of activities, but the old pros of the genetealogical world know that they are crucial to project momentum. Enthusiasm is contagious, so participant excitement is one of the keys to growing a project. And nothing pumps up a participant more than finally getting that long elusive answer or learning something he never knew before. Like it or not, it's often mundane analysis that produces the "Ahas!" that are the lifeblood of your project, especially in those cases where there's no match.

You'll also need to decide how to disseminate your results. Of course, you'll always inform the participating individual first, but more than likely, you'll want to share your findings more broadly, because these fresh details go a long way toward attracting new participants. Because this is such an important topic, we've devoted an entire chapter (Chapter 10, Interpreting and Sharing Results) to it.

8

Finding Prospects

When it comes to DNA studies, more is almost always better. The more participants, the better the chances are of linking previously unknown cousins, gaining insight into the origins of a surname, confirming or refuting traditional research, and solving age-old mysteries. And sometimes, only certain people have the right DNA to answer your questions. So how do you maximize participation and find the best candidates?

Two Approaches

As we mentioned in the previous chapter, you can recruit people in two ways—by finding them or by making it easier for them to find you. We refer to the detective work associated with seeking out appropriate candidates as "reverse genealogy" since it usually involves tracing lines from the past to the present. Traditionally, we're trained to start with ourselves and work back through the generations, but conducting a DNA project often requires the reverse. You may, for instance, be trying to find possible descendants of a German immigrant who came to Pennsylvania in the 1700s.

And we like to call techniques used to make it easier for would-be participants to find you "broadcasting"—essentially the equivalent of leaving a trail of crumbs on the Internet. Most administrators use both reverse genealogy and broadcasting strategies, but the specifics of your project will influence how you split your time between the two.

Reverse Genealogy vs. Broadcasting

As Figure 8-1 shows, the objective you set for your project will establish its scope, which in turn, drives who should participate—as well as whether it would be wiser for you to pour most of your energies into reverse genealogy or broadcasting efforts.

For example, your project may be narrow in scope (e.g., to determine if Jim Wolinsky and Bob Wolinsky, who found each other on the Internet, share a common ancestor). In this case, it's obvious that Jim and Bob are the ones who should participate, and there's no need to use any recruiting strategies. Or your project may be the opposite extreme—a broad study open to anyone with a particular surname, for example, to seek out any and all connec-

Figure 8-1: How scope affects participation and recruiting approach

Scope	Example	Participation
Narrow	Do Jim and Bob share a common ancestor?	Obvious participation
Midrange	Do the MA and VA Austins share a common ancestor?	Selective participation (mostly reverse genealogy)
Broad	Swanson surname	Open participation (mostly broadcast)

tions among people with the name of Swanson. Open projects such as this—especially ones focused on common names—will usually rely heavily on broadcasting tactics since it's not realistic for you to track each person down on an individual basis. Even so, there may be cases where you'll want to dabble in reverse genealogy to locate a particularly desirable participant, such as a descendant of a famous person who shares your surname.

But more and more projects fall in the middle range where participation is less clear (for example, to determine if various Massachusetts and Virginia Austin families share a common ancestor). In such cases, you'll want to use mostly reverse genealogy techniques to follow the Y-DNA trail—either to find the participants yourself or to qualify them once they contact you. Doing this is usually not too difficult because you're simply following the sons down through the generations—and with rare exceptions (for example, a known adoption), this means following a given surname. With mtDNA-based projects, the detective work becomes a little more challenging since you're following a maternal line and dealing with name changes each generation, but it's still very doable.

Since the majority of projects at present are broad—that is, open to anyone with a given surname—it's not surprising that the broadcasting technique currently enjoys the greatest popularity. But as more people launch midrange projects designed to answer specific questions, reverse genealogy is becoming a more important component of a coordinator's repertoire. Or maybe you're one of those who needs to find a proxy to test in your place—you're a woman interested in a surname project, or perhaps a man who wants to research one of his maternal lines. Whatever your situation and preference, the remainder of this chapter will provide plenty of ideas and guidance for locating as many appropriate participants as possible.

Reverse Genealogy: Following the DNA Trail

Locating DNA testing candidates gives you a chance to play sleuth and frequently rewards you with the discovery of previously unknown cousins as an added benefit. While it can be challenging at times, persistence and creativity in the hunt usually pay off. Walking through the process with a couple of fellow researchers will give you a feel for how to approach your DNA quest and what you might expect along the way, so we're going to take you step-by-step through a pair of actual searches here. We'll start first with a typical Y-DNA case and follow with an mtDNA situation, so you can view the process from both perspectives.

Y-DNA: Looking for Lucases

Stacy Gately was intrigued with the notion of genetic genealogy and had her father's Y-chromosome tested almost 2 years ago. The surprising results (a rare haplotype not matching anyone else of the same surname) from that first test whetted her appetite for more, so she decided to develop a *genetic pedigree*—that is, a Y- and mtDNA profile—for each branch of her family tree. She figured that by doing so, she would be well positioned to learn more as others got tested and DNA databases grew. She also thought that it would be smart to obtain samples while she still could—you never know when the death of a distant, last-in-line cousin might effectively end your chances to obtain DNA from that branch of the family. So Stacy set about trying to locate testing candidates for each name in her pedigree as far back as her great-grandparents' generation.

One of her great-grandmothers, Mary Lucas, was born in 1891 and immigrated to America in 1911. Mary had died long before Stacy was even born, but even if she had still been living, it would have been necessary to find an appropriate male relative to provide a Y-DNA sample. Since Mary was a woman, it was *her father's* Y-DNA that Stacy needed to trace. This meant that her true target—her *ancestor of interest* (see the top row in Figure 8-2)—was her great-great-grandfather, Phillip Lucas, and she needed to locate one or more of his direct-line male descendants.

Through traditional research, Stacy learned that Mary had been one of nine children, four boys and five girls. With so many potential lines to follow, she decided to sketch a descendancy

Figure 8-2: Following the Y-DNA trail— All of those surrounded by a dark, shaded area are testing candidates.

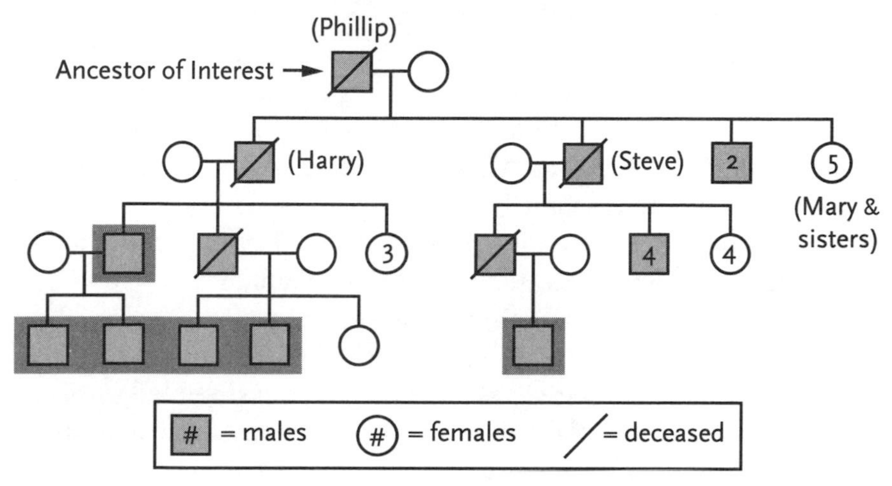

chart to help focus her search. To prevent the chart from becoming extremely busy, she detailed only the lines she intended to follow and collapsed the others by indicating how many other children of that gender there had been in the family (with numbers in the squares and circles). Phillip's five daughters would not have had the right DNA. After interviewing older relatives and learning that two of his sons had also come to the United States and settled in the Scranton, Pennsylvania, area, she opted to focus on them.

Armed with the names of the immigrant brothers, Stacy launched her reverse genealogy search with the 1930 census, the most recent one available to the public. This gave her approximate birth years as well as names of their wives and children— a good start! She now knew that Steve had been born around 1879 and Harry around 1897. While she could have followed either line, Stacy decided to trace both. If she found male descendants in both, she reasoned, she could get one from each tested and rule out the possibility of any non-paternity events.

Going straight to the Social Security Death Index, she quickly found a death date for Harry, the younger brother, but was not surprised that the one born in 1879 was not listed, since the bulk of SSDI records begin in 1962. To learn when Steve had died, she wrote to the county courthouse where he had lived and requested a copy of his will and other estate records. Now equipped with death dates for both, she obtained copies of their death certificates. Wanting to be absolutely sure she was following the correct DNA trail, she also requested their Social Security applications and naturalization records. All the documents confirmed that the brothers definitely shared the same parents as Stacy's great-grandmother Mary, and some records provided extra details, such as names and birth dates for some of their children.

With all the data she had gathered from the paper trail, Stacy hit the Internet. By searching the names of both of the brothers as well as their wives and children, she stumbled across the obituary of Harry's wife, which conveniently furnished the names of his now grown and married children as of 1992. Not having as much luck with Steve's family, she searched for newspapers in Scranton and found one with a Web site. A search of the newspaper's archives turned up a handful of articles with additional family details from the past decade or so, including a mention or two of the brothers' grandchildren.

Combining information from all these sources—census records, death certificates, naturalization records, Social Security applications, probate packets, and obituaries and other articles—Stacy was able to construct the third (and part of the fourth) generation of her descendancy chart (see the lower rows of Figure 8-2). Steve apparently had five sons and four daughters, while Harry had two sons and three daughters—meaning that there were plenty of Y-DNA lines to follow!

Stacy now turned to an online phone directory and began searching for the names in the most recent two generations of her chart. As expected, some of the people were not publicly listed, but several others were. She selected a daughter of Harry's and picked up the phone: "Hi, my name is Stacy. You don't know me, but I think we're cousins. My great-grandmother was your aunt Mary . . ."

An enjoyable half an hour on the phone later (those first calls

Candidate Tie-Breakers

In the Lucas case (see Figure 8-2), there are at least six candidates for testing. Should all six get tested? No. While it would be smart to get someone from both Harry's and Steve's lines tested to rule out non-paternity events, testing beyond that would be redundant, not to mention expensive! Barring any non-paternity events or recent mutations, the DNA test results of all of these men would be identical. In such situations, when you have more than one candidate, here are some tie-breakers for selecting the best one:

A descendant who bears the surname of interest (e.g., not his stepfather's), because it will be easier to explain results

The person who has the most close relatives, because that will provide a better cost-sharing opportunity

The person who's most interested and enthusiastic

can sometimes open the floodgates!), Stacy could flesh out her chart with more names as well as slashes to indicate those who had passed away. From Harry's line, she had five DNA candidates, and from Steve's, she had at least one for comparison's sake, with the prospect of four other lines that could be pursued if necessary. She also had the pleasure of making the acquaintance of a delightful first cousin twice removed and an invitation to visit and see photos of her ancestor of interest, her great-great-grandfather Phillip Lucas—a nice bonus for her sleuthing efforts!

mtDNA: A Soldier's Tale

Now let's shift gears slightly and look at an mtDNA case that one of us researched for the U.S. Army. African-American soldier Cleveland Payne was born in 1912 in Illinois and lost his life in Korea in 1950. The Army wanted to find his family today, including relatives who could potentially provide an mtDNA sample to compare against any remains that might be located. Due to a fire in 1973, Cleveland's personnel record had been destroyed, so details were skimpy.

The hunt began with the 1920 and 1930 census in order to learn the names of Cleveland's parents and siblings. Since he had obviously received his mtDNA from his mother, she was the ancestor of interest. Fortunately, Cleveland had come from a large family and had four brothers and five sisters, so there were many leads to follow.

Because the name Payne is relatively common, however, a search of the SSDI turned up only one individual who could be clearly identified as a brother. With no other obvious options, the brother's death certificate was ordered. But the document led to another roadblock because the brother had apparently died unmarried and without children, and the informant was a hospital employee, rather than a relative. Luckily, the certificate yielded another clue—the name of the funeral home that had handled the burial. A call to the funeral parlor resulted in the married names of three sisters who survived their brother as of 1977—a significant leap forward!

Returning to the SSDI with the sisters' names revealed that at least two of the three had since died. A check of online news-

Figure 8-3: Following the mtDNA trail—All of those shaded, without slashes, are testing candidates.

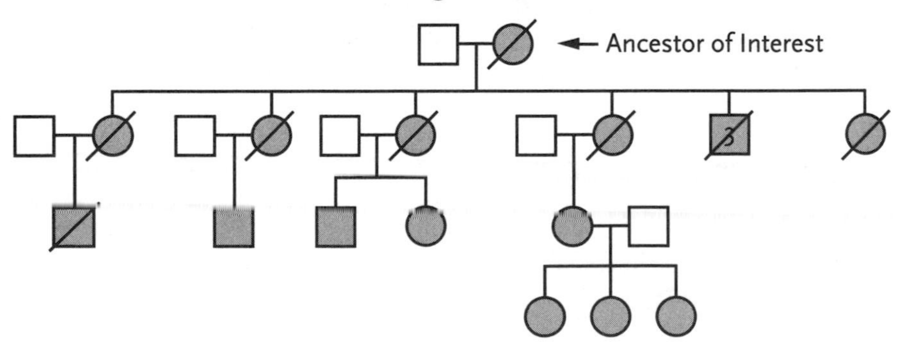

paper archives quickly produced the full text of both of their obituaries, but one proved to be another dead end because the sister had only had one child, who had predeceased her, leaving no mtDNA lines to follow. The obituary for the other sister was only slightly more helpful. She apparently had a daughter, but the daughter had married into the name of Smith, meaning that she could be very difficult to pinpoint.

Once again, a funeral home (mentioned in the obituary) moved the case forward by providing contact information for the Smith daughter as of 1996, but the phone number had been disconnected. A reverse lookup of the address at an online phone directory produced dozens of hits, suggesting an apartment complex. Scanning them, a number for the management company was located. The person who answered the phone indicated that Ms. Smith had moved 2 years ago but still lived in the area. She agreed to take the Army's contact information and ask around for this niece of the soldier.

The niece mustn't have moved far because she called the next day and provided the names of seven mtDNA-eligible people in the family. Typical of mtDNA cases and our society today, the seven candidates now sported six different surnames through marriage (none of which was Payne), and the family once situated in Illinois was now scattered in Oregon, California, and Arizona. And while Cleveland Payne's story is still incomplete, it's hoped that he, like some of his compatriots who served in Korea more than 50 years ago, will finally be identified and properly interred with the honors and recognition due for their sacrifice.

As you've probably realized, mtDNA searches tend to proceed much like Y-DNA ones, except for the additional challenge of dealing with name changes each generation. For this reason, they usually involve heavier reliance on documents that might divulge women's married names, such as obituaries, wills, marriage records (with brides' indexes), and death certificates. But the same reverse genealogy approach can be used with both Y- and mtDNA situations.

Reverse Genealogy Guidelines

If you have reason to use reverse genealogy to identify and locate potential participants for your study, a few straightforward guidelines can make your search much easier. Fortunately, we've had lots of experience in this area, so we'd like to share some tactics we've found to be especially helpful. We have deliberately emphasized examples from the past 50 to 150 years (because many genetealogists will find themselves focusing on this timeframe) and widely available resources (such as those found online), but depending on your specific circumstances, you may find yourself dealing with earlier centuries and more traditional, non-Internet-based research. In either case, the following general principles still apply.

Remember to Surround and Conquer

Back in Chapter 1, If You're New to Genealogy, we suggested a surround-and-conquer approach to research, and that definitely applies when it comes to finding DNA testing candidates. Rather than fixating on one individual and allowing him or her to become a bottleneck, your search will often be more productive if you expand your scope to include relatives, friends, and other associates. You can use the information found this way to work your way back to the targeted person and then move forward in time to his or her descendants, your potential testing partners. Doing so, for instance, may reveal a previously unknown married name for a woman whose mtDNA trail you wish to follow—perhaps in her father's Civil War pension file, her brother's will, or her mother's obituary. This, in turn, would allow you to find records pertaining directly to her that would likely mention her children. It may seem like a detour initially, but you'd be surprised how much time these seemingly indirect routes can save you! And incidentally, the closer you get to the present, the more useful you're apt to find this approach. The living are well protected by privacy laws, unlisted numbers, and other mechanisms, so reaching them through associates—perhaps military or school buddies, neighbors, or fellow members of their sailing or antique car club—may be your most efficient means of contact.

Remember the Women!

Since the majority of genetic genealogy that's being conducted today is focused on surname projects, this seemingly cuts women out because they don't have a Y-chromosome. But bearing the surround-and-conquer tactic in mind, it's important to remember that women can participate by proxy. If they're interested in learning about their maiden name, for example, they can talk a brother, cousin, father, or uncle into testing in their place.

Since genealogists are apt to be more interested in DNA projects than people contacted at random, it's useful to know that somewhere between 63 and 72 percent (depending on which survey you choose to believe) of genealogists are female. This means that anyone considering launching a DNA project would be smart to make an effort to enlist the help of women in recruiting related men to participate. In fact, it was interesting to learn from a survey we conducted of DNA projects that approximately a third of study administrators are women and that women are far more likely to run multiple surname projects than men. They may not have the Y, but they sure know how to find it!

Choose Your Initial Target Wisely

If you're fortunate enough to have several names you could pursue—say, a cluster of siblings born to a couple named Brown between 1850 and 1870—begin your search by focusing on these:

1. the most recently born
2. the one with the most unusual name
3. a male

The youngest person is your closest bridge to living descendants. If you start with the child born in 1870, you're essentially already a generation closer to today than if you had opted to start with the one born in 1850. And if the children are named Thomas, John, James, Elijah, Mary, and Anne, you're apt to have more success looking for Elijah Brown than the others, whose names are painfully common. Finally, although we don't wish to be sexist, it is true that males are often easier to follow through the generations because they retain their surnames, so all other factors being equal, you'll probably save time by starting with a man (unless you're tracing an mtDNA line that makes a woman your primary focus). If your efforts to find the first person fail to produce results, apply the surround-and-conquer axiom above, and move on to the others.

When Necessary, Go Backward to Come Forward

Most of us today are trained to think in a linear fashion. When we're researching our family trees, we start with ourselves and methodically work our way back through the generations, so when we're doing reverse genealogy, we expect to start at a given point in the past and steadily march our way forward in time. In principle, this makes a lot of sense, but based on our experience with hundreds of reverse genealogy scenarios, we've found that a zigzag pattern—going back and forth through time—is often more efficient. Partly, this is because taking a step back in time often allows you to gather the names of relatives and associates—yes, surround and conquer yet again!

For instance, we had one case where we found a particular family in the 1930 census, but couldn't find traces in more recent documents. So we backed up to the 1900 census to find the father of the 1930 family as a child with his parents and siblings. Fortunately, one of his brothers had an unusual name, and we were able to leap forward to 1969, the year this brother died and was listed in the Social Security Death Index. We went into reverse once again, locating this brother in the 1930 census with his wife and children. Finally, we combined information from the SSDI (where he died) with data from the 1930 census (the names of his children) and searched current phone directories for his now-grown offspring in that town. Bingo! It was a simple matter of asking the gentleman we located about his first cousins to find the family we initially sought. We were blocked when we tried to go straight from 1930 to today, but a zigzag 1930–1900–1969–1930–2003 pattern did the job.

In the course of research done to locate DNA candidates, you may also occasionally find it necessary to shift into reverse because the lines you're pursuing have died out in terms of the type of DNA you're seeking. (This is sometimes referred to as *daughtering out* if you're seeking a Y-DNA line and a few have suggested *petering out* as the mtDNA equivalent.) Perhaps you're trying to follow a Y-chromosome line, for instance, and the last known individual in the family has died childless. In this case, you would need to back up a generation to look for other lines to potentially follow into the present. Perhaps his father had brothers whose descendants could be traced, but if not, maybe his grandfather did. You may well discover that sometimes the only way to move your research forward is to go back in time first!

Follow the Trail of the Deceased to Find the Living

Those who have already tried their hand at locating DNA testing candidates often discover that the last step is the hardest. Many a researcher has found someone listed as a youngster in the 1930 census, but not been able to find that person today. One reason is our geographic mobility, and we don't mean just recent decades. When trying to locate families of soldiers who served in WWII and Korea, for example, we find a disproportionate number of relatives in California and Northwestern states due at least partly to the lingering effects of Depression-era migration.

Again, thanks to privacy laws and the proliferation of unlisted and cellular phone numbers, it can be very challenging to make the leap from 1930 to the present. Over the past few years, many states have enacted stricter privacy laws pertaining to records of the deceased, but even so, it's usually easier to obtain data about those who have passed away than the living. For that reason, we suggest making heavy use of resources such as the Social Security Death Index, online state death indexes, and obituaries. The SSDI is one of the most helpful resources for discovering where a family you're seeking may have resided within the past few decades, and if a state index exists, you may be able to secure a few extra details. Obituaries are even better because they frequently list survivors and where they reside. So if you can't find a particular person, it's smart strategy to look for the paper trail generated by the deaths of their parents, spouses, and siblings.

Best Resources for Reverse Genealogy

Now that you're equipped with some reverse genealogy tactics, the question that naturally springs to mind is where to start. If you've been doing genealogical research for any period of time, you're familiar with the standards, such as www.ancestry.com, www.familysearch.org, www.genealogy.com, www.rootsweb.com, and www.cyndislist.com. These are almost certainly among your bookmarks. But if your goal is to try to locate a DNA testing participant, which resources will help you find him the quickest?

Through our extensive tracing experience, we have learned that no two situations are alike, so the resources used may vary widely. Still, we realized there were certain "old reliables" we turned to over and over in these quests, but we were hard pressed to rank the top performers—the ones that were the most essential to success. So we decided to conduct an experiment.

We randomly selected 10 cases we had worked on and retraced the research trail, logging the resources that had been used in each one. All cases involved the use of several resources, and a given tool may have been consulted multiple times. To simplify matters, we counted a resource only once for each case in which it contributed to the solution. A convenient ranking quickly emerged, as can be seen in Figure 8-4.

Every-Name Census

The clear-cut superstar is the digitized, 1930 every-name census index (available for a fee at www.ancestry.com), with the 1880 (available at www.familysearch.org and www.ancestry.com) playing a supporting role. This tool was a key ingredient in 80 percent of the cases we analyzed.

With the ability to use multiple variables (state, age, place of birth, etc.) as well as wildcard spellings to zero in on your target—one among its 124 million entries—the 1930 census is an indispensable resource. And since everyone in the family is listed, it's a potent surround-and-conquer weapon for finding additional names to pursue. Of course, its true value, as alluded to earlier, stems from its recency. When you find a name here—particularly a child—you're already dealing with someone who may well be alive. And if they are deceased,

Figure 8-4: Best reverse genealogy resources

Rank	Resource	Frequency of Use (%)
1	Every-name census (1930, 1880, 1870, 1860)	80
2	Online lineage collections	70
3	Online phone directories	60
4	Social Security Death Index	50
5	Online State Vital Records	40
6	Other census indexes (e.g., 1900, 1910, 1920, etc.)	30
7	Search engine (e.g., Google, etc.)	20
8	Other sources (e.g., newspapers, real estate, etc.)	20

odds are that they lived long enough to leave a trace in the SSDI.

The 1880 census is also very powerful due to the fact that all 51 million entries have been indexed, rather than just the heads-of-household as is the case with most census records. In fact, the 1880 and 1930 censuses often work well in tandem. If you can find someone as a child in the 1880 census, you can often find him as a 50-something in the 1930 census, enabling you to almost instantly leap half a century forward in time. Recently, the 1870 and 1860 censuses were also completely indexed and may prove almost as useful.

Online Lineage Collections

The second most useful resource, which contributed to 70 percent of the cases, is the ever-expanding collection of online family trees uploaded to the Internet by your fellow researchers. By these, we mean the ones that can be found in such places as Ancestry World Tree (www.ancestry.com), World Family Tree (www.genealogy.com), and Ancestral and Pedigree Resource files (www.familysearch.org). We were actually surprised to find this resource so highly ranked, especially given the fact that it's no secret that there's plenty of misinformation in online lineages. Taking a closer look, we discovered that there's usually enough correct information about recent generations—the ones that are often toughest to trace—to lead you to someone in the family today. The number and mushrooming content of online trees means that many of us will find some potential leads embedded in them, and making contact is easy since you can simply e-mail the individual who submitted the tree that interests you.

Online Phone Directories

Online phone directories often provide the last bit of information necessary to make contact, so it's not surprising that they placed third in importance. Since approximately one-third of Americans have unlisted numbers, it makes sense that 60 percent of cases were facilitated with this resource. Our favorites change over time due to shifting functionality, but our current default is www.whitepages.com, which incorporates handy features such as reverse lookups on phone numbers and addresses. If we choose to search several directories at once, www.theultimates.com gets our vote, and for international searches, www.infobel.com/teldir provides a valuable launching pad with its links to directories for countries around the globe.

Social Security Death Index

We've already mentioned how useful the SSDI is, so if anything, we might have expected it to place higher than fourth. Many people found in the 1930 census can be traced forward in time by locating their SSDI entry, and the locations mentioned for last residence and benefit can point you to an area where you will find family members today. This information, in turn, can steer you to other resources, such as local newspapers that may have obituaries.

The SSDI is available for free at several sites on the Internet, and like phone directories, their functionality may be tweaked from time to time. These days, we tend to do most of our sleuthing at www.rootsweb.com and www.familysearch.org, two sites that offer complementary search flexibility (e.g., wildcards at Rootsweb and name variations at FamilySearch). We've also found Rootsweb to usually be the most up-to-date.

Online State Vital Records

The past few years have been tough ones for genealogists in terms of access to online vital records indexes. For both legislative and fiscal reasons, many states have taken measures to restrict access to such databases (and the documents they cover), making our research more challenging, but there's still more out there than many realize. As of this writing, for instance, a subscription to Ancestry.com provides at least some 20th-century indexes for 24 states. A few states, perhaps recognizing the income-generating potential of genealogical orders, have uploaded their own indexes.

Illinois, for instance, allows you to search for deaths that occurred from 1916 to 1950. Since the availability of these resources varies widely by state and is in such a constant state of flux, you may wish to bookmark Joe Beine's *Online Searchable Death Indexes* (www.deathindexes.com) to keep up to date. And if you can't find anything on the Internet, it's always worth searching the Family History Library Catalog (www.familysearch.org) by state and county (look under the "vital records" category) to see if perhaps the library has any relevant microfilms.

The value of these fifth-place records is somewhat similar to the SSDI (e.g., narrowing the date of death to make it easier to find an obituary), but you may be able to obtain extra details if you qualify to obtain a copy of the relevant certificate from the state in question. For instance, a death certificate may give you the name of the informant, often a surviving spouse or child, or perhaps lead you to the cemetery where you can find other family members buried in the same plot.

Other Census Indexes

Coming in at sixth place are other online census indexes—that is, the ones that have been indexed primarily by head-of-household, rather than by every name. At present, we tend to search the 1920 at www.ancestry.com, the 1910 at www.nygbs.org or www.godfrey.org, and the 1900 at www.godfrey.org or www.genealogy.com. All of these are fee-based but reward you with digitized images when you find a hit. Searching each of these, rather than jumping from 1880 to 1930, can help you flesh out the family tree and obtain additional names (e.g., a daughter born in 1891 who married before 1910 will appear with her birth family only in 1900). Earlier census years are also available at all three of these sites.

Search Engines

When confronted with a situation where we have the name and place of residence of someone living today but cannot find an address or phone number, we often turn to search engines. Of course, you can try your favorite engine at any point in a candidate-hunt (especially if you're fortunate enough to be dealing with an unusual name), but we find them especially helpful for this particular roadblock.

In one recent effort, for instance, we had a name and town but came up empty in the phone directories. A quick visit to www.google.com to enter these details produced an article about a policewoman in a local newspaper. She was the one we were looking for, and the article furnished enough information to make the final contact possible. The effectiveness of the name-location combination in circumventing the unlisted number phenomenon earned search engines the seventh position in our experiment.

Other Sources

Holding up the rear is a collection of resources that may be used for specific circumstances. This catch-all category includes newspapers, obituary and cemetery resources, county-based Web sites, public records (e.g., real estate, incorporations, licenses, etc.), and specialized sites, such as those geared toward military personnel and school buddies. Unfortunately, discussing all of these would easily fill another chapter, but we would like to mention that newspapers—loaded as they are with obituaries, marriage and birth announcements, business dealings, and sometimes amazingly trivial tidbits about your ancestors—are coming on especially strong with various vendors offering a growing number of digitized and fully searchable collections. In fact, we suspect that if we were to repeat our experiment in a year, they would emerge as their own category. Among our bookmarks for newspapers are www.newslibrary.com, www.newslink.org, www.ancestry.com (the Historical Newspapers collection), www.godfrey.org (for *The New York Times* and other major newspapers), and www.obitsarchive.com.

Broadcasting

The preceding were our recommended guidelines and resources for reverse genealogy, but what if you've decided to launch a broad project, open to anyone with a given surname? In that case, you're probably going to adopt a broadcasting strategy, so let's start by peeking over the shoulders of a few people who made the same decision.

Broadcasting: Rose

David W. Brown, who manages the Rose DNA Study, one of the largest DNA surname projects in the world with more than 140 participants, believes that it is "almost essential to have a surname publication that is widespread—at least to grow to the size of our project." He explains that having a dedicated genealogist on board is a significant advantage and that many of their participants are now recruited through the efforts of Christine Rose, a well-known and highly respected professional who also serves as the primary researcher for The Rose Family Association. In addition to getting the word out through the *Rose Family Bulletin*, the project maintains a Web site (www.ourworld.cs.com/Christine4Rose/) and taps into other resources, such as postings on various genealogy boards, their testing company's surname list, and word of mouth.

Broadcasting: Hull

James Reynolds Hull, Ph.D., confesses to "organizational jitters" when initiating the Hull Surname DNA Study. He started by sending 500 letters of introduction, an effort which was not very successful. Fortunately, as a long-time member of the Hull Family Association (HFA), he realized that he had a built-in database of potential Hull donors. Working together with the HFA and its genealogist, he has been able to seek and find those Hull men who are known to be related by way of tradition or historical documentation to Hull progenitors.

Broadcasting: McCarthy

Perhaps one of the most ambitious outreach efforts can be credited to The McCarthy Surname Study, run by Cliff McCarthy and Barbara McCarthy. Open to all males with variations of the McCarthy name (e.g., McCarty, MacCarthy, etc.), the project administrators began seeking participants by placing announcements on genealogical message and mailing boards and having the study added to master Y-chromosome project listings. Capitalizing on the established reputation of The Clan MacCarthy Society, they then sent letters to all members and had information posted on the Society's Web site. Going a step further, they also did a direct mailing to all 43 McCarthy males in the Dunmanway area of County Cork, Ireland (a region of particular interest), and sent flyers to the Dunmanway librarian requesting that they be posted on local bulletin boards. Yet another tactic used was contacting relevant organizations and periodicals in both the United States and Ireland, asking that a write-up about the study be printed in their publications. Last but not least, Cliff and Barbara relied on the power of word of mouth to attract more candidates.

Broadcasting Guidelines

You may not be planning as large a project as the Rose, Hull, and McCarthy studies, but we hope in reading about their experiences, you've found a few ideas worth mimicking. And while the broadcasting approach is more straightforward than reverse genealogy, keeping a few guidelines in mind can help those crumbs you sprinkle lead more testing candidates your way.

Leave Lots of Traces

We'll discuss the best resources for getting the word out shortly, but the primary objective with the broadcasting strategy is making yourself findable, so you'll want to toss out as many crumbs as possible. And just as with advertising, there's power in repetition. A single announcement on a surname message board may net you several inquiries, but popping up from time to time will help your project gather more momentum.

Follow the Rules

Having just recommended that you post periodically, we now need to offer a qualifier to this advice: Follow the rules to the n*th* degree. Some board and mailing list administrators are less than receptive to DNA testing. They may not care for it personally, or they may regard messages about projects as nothing more than crudely disguised sales pitches. In a few extreme cases, administrators have even been known to routinely delete any postings referencing DNA. Rather than do battle with such people, experienced DNA project managers have found it more effective to do all they can to accommodate them.

Make it a habit to read the rules before making your first posting to any board or list, and never include dollar amounts. For that matter, don't make any mention of money or payment, even though you may wish to do so in the interest of full disclosure. The appearance of commercialism is the most frequent reason given for posting refusals. And while repetition is important, don't wear out your welcome by simply announcing your project over and over. After the first message or two, add value to your announcements by sharing results or insights that will be of interest to the other readers. Stick to the relevant topic, and then direct people to your Web site or e-mail address for additional details. Finally, if you want to maximize the chances of all your messages appearing, you may even wish to engage in a private e-mail exchange with the administrator before making your first posting. Such consideration can go a long way to minimizing potential conflicts down the road.

Do a Little Harvesting

Just as you're leaving a trail on the Internet, others have done so before you—and frequently with their e-mail address included. This means that if you look in likely places, such as relevant surname and locality boards, you'll be able to find contact information for people who might be interested in your project. Mind you, we're not advocating any sort of automated approach where you collect e-mail addresses of anyone with your surname, but we feel that sending a message or two to individuals who have posted messages on pertinent genealogical resources is reasonable.

And while most who use this tactic look for candidates specifically on boards and mailing lists, we suggest that you expand your reach by searching online lineage collections as well. Researchers have different preferences; some who like to upload pedigrees may not participate in surname boards, so you may find extra candidates this way. Also, if you're seeking particular testees—say, Irish-born Reynolds—these online trees often provide enough detail to enable you to hone in on such people. A search in Ancestry's World Tree collection, for instance, turns up more than a quarter of a million Reynolds entries, but restricting it to those born in Ireland brings the number down to less than a thousand—many of which were submitted by a small cluster of individuals whose e-mail addresses appear along with their data. Of course, there will be considerable overlap in the entries, but these online lineage collections provide an often overlooked means of learning about fellow researchers who are interested in a given name.

Create a Web Site

This is not an absolute requirement, but one feature all the largest DNA projects have in common is a Web site. And while creating the site may require an investment of time, it can ultimately save you much more by allowing you to share everything once rather than in a constant series of one-on-one communications. Having a strong Web site gives you an obvious place to direct board and list readers for more information. It allows you to cover the basics of DNA testing, so you won't have to coach every would-be testee individually. It permits you to easily share results so participants don't have to rely solely on you for their information. It can extricate you from the middleman role by providing contact information for others involved (with their permission, naturally). And of course, it will help you attract more participants by building credibility and always being there when someone has a question. We'll consider the ingredients of an effective Web site and share some of our favorites in Chapter 10.

What's the WIIFM?

If you're trying to recruit participants, it always helps to put yourself in their shoes to try to understand what they might hope to derive from testing. Addressing the WIIFM—the "What's in it for me?" question—might lead to some creative ideas. For instance, many people have a family tale of being related to a famous person with their surname and would like to know if it's really true. If you can locate an appropriate relative of the famous individual (maybe you can't reach close relatives, but second, third, and more distant cousins may be accessible) and persuade them to join your study, others would then have a means to see if they truly do share a common ancestor with the famous person. An example of this can be seen in David Roper's project focusing on Ben Franklin's DNA (www.roperld.com/FranklinBenDNA.htm), although we want to be quick to point out that David has long researched many Franklin lines.

As another example, many Irish genealogists are stymied in trying to ascertain where in Ireland their ancestors came from, so you could conceivably contact people with a given name still living in Ireland today and request that they participate. Once several agree, those of Irish descent from other countries might be more interested in getting tested in the hope of matching a present-day resident, thereby providing a hint of where they should focus their future research in Irish records. Even the prospect of a match would encourage some to participate.

To get the creative juices flowing, just ask yourself what you would hope to learn under ideal circumstances from participating in your own project. Figure out what your dream scenario is, and see if there are any steps you could take to improve the likelihood of it actually happening for you or anyone else in your study.

Broadcasting Resources

The DNA project manager has many broadcasting options, but as administrators ourselves, we were curious to know which ones were the most popular and the most effective. So we included a few questions on this topic in our survey of experienced coordinators. We've distilled the results in Figure 8-5. The popularity of a given technique is measured as a percentage of projects using it (and rounded to the nearest percent), while effectiveness is admittedly somewhat more subjective. Based on our interpretation of managers' comments, we rated each tech-

Figure 8-5: Popularity and effectiveness of broadcasting techniques

Technique	Popularity (%)	Effectiveness
Surname and locality-based message boards and mailing lists	68	high
Via family/surname association (and its newsletter and/or membership list)	33	high
Web site	30	moderate
E-mail (mass or list mailings)	30	moderate
Testing company lists	13	moderate
Traditional mail	18	slight
Word of mouth	15	high
Phone calls	13	high
Master DNA project lists (e.g., www.dnalist.net)	5	moderate
Reunions	5	high
Announcements in genealogical publications	3	N/A
Other (e.g., newspaper articles)	3	N/A

nique as highly, moderately, or slightly effective. The results are shown in descending order of popularity.

Not surprisingly, most project managers use a combination of techniques to improve their results, with many claiming four or five approaches. And we had anticipated finding good old-fashioned message posting leading the way! Beyond that, we were mostly interested in the disparities between popularity and effectiveness. For instance, almost 20 percent had tried snail mail (more than we would have imagined) but found it to be disappointing.

But the reverse is that phone calls and reunions are tactics that apparently offer untapped opportunity to recruit new participants. Jim Hull, whom we encountered a short while ago, runs several DNA projects and is convinced that personal telephone calls recruit the most people: "Since it is 'now' communication, the questions get answered, and I can get some sense of why a person might be hesitant and respond to those concerns."

And Mary Lou Hudson has had great success using family reunions to get people interested in DNA testing for several names: "My Cox reunion in Texas budgets money each year for DNA testing. This past year, I made a motion at my Creekmore reunion to sponsor testing, and it was accepted. After I returned home, I received several messages from men who attended and wanted to be donors. Last year at my Anderson reunion, the group voted to sponsor DNA testing, and I asked again this year for them to support more testing. So far, the reunions I am involved in have been more than happy to get involved. If a group holds family reunions, that would be the time to get people interested in DNA testing."

We also noted that word of mouth is another underutilized technique, and since "happy customers" are the best proponents, now seems a good time to turn our attention to the people side of participant recruitment.

9

Contacting and Courting Participants

If you've used reverse genealogy tactics to find a candidate, you've got a potentially delicate situation on your hands. After all, you're about to ask a stranger for a sample of his or her DNA! And even if they've found you through your broadcasting efforts, they will almost certainly have some questions and concerns. In our experience, how you handle these interactions is critical to the progress of your study.

Excuse Me, May I Borrow Some DNA?

We're not going to sugarcoat it. Talking strangers into handing over their DNA—and hopefully, some money—is not the easiest of tasks. Presumably, it will become easier over time as genetic genealogy becomes as widely known as traditional research. At least then, those you contact will know that this is a normal activity that everyday human beings do with some regularity, and there will no longer be a need to educate people about the very existence of this kind of testing. But it's best to prepare as if the person you're about to call, write, or e-mail has never heard of genetealogy.

It's All about Trust

The key to securing participation is trust. People must trust you if they're going to cooperate. Many active genealogists already belong to or run family associations of some sort. Perhaps you produce the family newsletter. If this is the case with you, your job just got a lot easier because those face-to-face meetings at reunions or newsletters consistently showing up in members' mailboxes will make you familiar to them. When one of us launched her first project, she immediately had half a dozen self-funding participants (when testing was more expensive) because candidates from her decade-old village association already knew and trusted her.

If you're contacting complete strangers, patience is the not-so-secret ingredient that will eventually win you trust. Hit-and-run approaches "Here's what I'm doing. May I have your DNA, please?" will not work. Some of you may have had the experience in your traditional research of cold-calling distant relatives who have never heard of you. If so, the same principles apply. You'll want to identify yourself, and if phoning, ring in under your own name if possible (those of you who are unlisted won't be able to). This will get you past some call screeners and also reassure people that you are who you claim to be. If you have an unlisted number, you may have to be a little more persistent.

And you'll want to establish your bona fides. If you're a distant relative, it helps to quickly drop some family tidbits into the conversation—preferably details that only family members would know. If you share a mutual great-great-grandfather, for instance, mention that he was the immigrant who came from England and settled in Cleveland, Ohio. If you're lucky enough to have reached a fellow genealogist, you'll probably find things flowing smoothly from this point with the bulk of their questions centering on DNA. And even if you've reached someone who's just vaguely aware of the clan's history, you'll pique their curiosity and earn a little trust by sharing information they know to be true before asking for any in return.

If you're contacting someone with no known relationship—say, for a broad surname study without a supporting family association—it may be a little more challenging for you. Those with particularly unusual names may be intrigued to hear from someone else with the same name and will usually give you at least a few minutes of their time to hear you out or read what you've written, but those with relatively common names are less likely to have this advantage.

The Initial Contact

So what do you talk about in that first phone call, letter, or e-mail? After explaining who you are and briefly covering your research efforts, tell them what you're trying to accomplish. Acknowledge that it may sound a little strange, but you're contacting them to see if they would be interested in participating in your study—and resist the urge to directly ask for a DNA sample. In fact, if you can discipline yourself to wait, steer them to your Web site, or offer to send additional information (including perhaps referrals to others who are in your study), and say you will contact them again in a few weeks once they've had a little time to absorb it all. Your chances of success improve immeasurably if you don't aim to gain participation on the very first contact.

Then prepare yourself for an onslaught of questions! Some will pummel you with queries immediately, and that's usually a good sign. They're at least curious about your project and are not dismissing it out of hand. Others will want to "see something in writing" or learn more about the subject before moving forward. In either case, do not be alarmed if you find yourself going through several cycles of answering the same questions. In fact, expect it. You're reading a book on the topic and may already have some actual experience, so it's easy to forget how startling, intimidating, or confusing the world of DNA testing may have seemed to you at first. The folks you're contacting are more than likely in that novice mode, so it's your job to educate them if you want them to join your study.

Questions and Misconceptions

What kinds of questions can you expect? Many of them will center on your motivation and the testing process itself, so it helps if you have a tidy and sincere sound-bite to explain the objective of your project, as well as copies of and/or links to basic articles on genetealogy to help bring others up to speed. Many project managers have Web sites with Frequently Asked Questions (FAQs) sections or links to other administrators' FAQs (such as the *DNA 101* section on the Blair DNA Project Web site: www.blairgenealogy.com/dna/dna101.html), and these can spare you some of the redundancy in the question-and-answer cycle.

Since you've already established your objective, and much of this book is devoted to the testing process, we thought it would be most helpful to discuss the resistance you might encounter, so you'll be better equipped to deal with it. But first, let's take a brief detour to point out two common misunderstandings pertaining to the tests themselves. While they don't materialize in the form of resistance, they can create a little consternation or confusion, so it's best to address them early and head-on.

The first is the belief that women can participate in Y-DNA projects. As we've discussed, they can, but only by proxy—by recruiting a brother, cousin, uncle, or other male relative to get tested in their place. While most people quickly grasp this concept, every once in a while, you may encounter a woman who does not appreciate that the reason she can't get tested for a surname project is scientifically rather than chauvinistically based. "Why can't I get tested?" you'll hear in protest. "I've got DNA too!" Well, yes, she does, but not the right kind for this particular test. A good way to avoid this scenario is to include it as a FAQ on your Web site or in any materials you share with would-be participants.

The second common misconception? That testing will magically reveal one's entire family tree or very specific information, such as the name of a common ancestor. The same folks who think they can do a vanity search on the Internet and instantly find their pedigree are the ones who think one test will tell them everything they ever wanted to know about their family's past. As we've discussed in earlier chapters, both Y and mtDNA testing can only shed light on a single branch of the family tree—and even that doesn't come with names and dates attached. Fortunately, many people are most interested in one of these two branches in any case, so this reality is rarely a show stopper, but it's smart to make sure that all participants—even those who are gung ho to get tested—understand this up-front. This little bit of insurance will prevent possible disappointment when the test results arrive. In fact, many project managers also devote a FAQ to this matter as well.

Resistance

Brace yourself! Not everyone is going to be salivating at the chance to provide you a sample of their DNA (pun avoidable, but too tempting!). In fact, some may have not only questions, but objections—reasons why they're reluctant to participate. On the theory that knowing them in advance will help you cope when they crop up, we asked veteran project managers about this issue, blended in our own experience, and summarized the most frequent stumbling blocks:

Scam, Indifference, and Not-Real Genealogy

This is our catch-all category for people you might want to consider leaving out of your study. Some inherently suspicious people will accuse you of trying to pull a scam on them. The good news is that such people are usually quite vocal, so you'll know exactly what you're dealing with. While you might be able to slowly gain their confidence with persistent, gentle communication, experience shows that these folks are rarely won over.

Indifference is harder to spot because it's often passive. People may express a willingness to participate, but never order a test kit—or even more frustratingly, order a kit, but never return it. Again, follow-up sometimes helps (as do offers to share the cost, but more on this later), but after a certain period of time, you're probably better off devoting your effort to recruiting others.

Finally, because genetealogy is still relatively new, there are some naysayers within the genealogical community itself. Some believe that those doing genetic research are trying to use it as a substitute for traditional research or that we're not aware of its limitations. For such reasons, they'll decline to participate in DNA projects. In their view, this is likely a matter of principle, which makes it particularly difficult to alter their opinion. For this reason, we suggest that you also forego such participants. They may shift their view over time, but it's not likely your persuasion will be the catalyst for this change.

Of course, the exception to all of the above may be if that individual's participation is crucial to your study. Perhaps they're the only person with the "right" DNA to solve some centuries-old mystery. In that case, persistence may be warranted, but if they're not essential to your project, we recommend moving on to the next person when you encounter any of this troika.

Confusion about the Test

In addition to the misunderstandings we mentioned previously about whether women can participate and how much of the family tree is covered by testing, there is often confusion between DNA testing done for genealogical reasons and that done for medical or criminal purposes. Some are under the impression that their results will reveal medical information, such as the predisposition to certain diseases, and will be concerned that insurance companies could use it against them in the future. By contrast, others will be disappointed to learn that they won't obtain any medically useful insights from the test. Similarly, you may encounter a few who fear their test results could wind up in a criminal database or be used to implicate them in a crime. In all cases, you'll simply need to explain that this type of testing differs (as discussed in Chapters 3 and 4) and that neither medical nor criminal information will be revealed.

Privacy Concerns

This is perhaps the single biggest issue of potential participants and should not be taken lightly. Many have reasonable "Big Brother" apprehensions, and why shouldn't they when DNA is touted on a daily basis as a unique identifier and the code to all human life? Could this data somehow wind up in the wrong hands? Could their participation in a study ever come back to haunt them in some way?

Fortunately, the project manager has a great deal of control over this matter and can take steps to protect everyone's privacy. If you spend a few minutes browsing the Web sites of the more established DNA projects, you will frequently see confidentiality policies spelled out (and sometimes included in the consent forms discussed earlier). Some projects are kept entirely private with no results shared except with the participants themselves, but that's already the exception, rather than the rule. Most coordinators either limit access to results (via a password-protected Web site) or only post results under codes rather than names (such as Smo1 for a particular Smolenyak participant) or the name of the earliest known ancestor (such as Daniel Shields, born 1741 in Ireland), so that no one's privacy is compromised. This last is rapidly becoming the preferred approach because it grants the participants anonymity while still allowing others to see the results and determine if their Y-DNA is represented because they share an ancestor with someone who has already been tested.

It should also be pointed out that it's not cause for serious alarm if someone's name and markers happened to be posted publicly someplace. Because genealogical DNA testing unveils such a tiny fragment of our genetic makeup, the most anyone could do with the information is see if they match the individual or perhaps determine if he's from a given haplogroup. But in spite of the very limited potential for any sort of damage stemming from the public sharing of this information, it is best to err on the side of safety, develop a confidentiality policy, and then rigidly adhere to it.

Blood and Needles

Although many now know that most DNA testing is done with the use of swabs, there are those who still think it's done through blood tests. And lots of people hate needles! One major study that used blood tests had to develop other collection methods for a variety of reasons, one being that their database was becoming lopsided with far more females than males. Why? Because more men are afraid of needles. If blood tests were our only option, most surname-based projects might be considerably less populated!

Happily, this is another easy area to address because the test is simple and painless. Should you encounter any people who have qualms about the DNA collection process (more common than you might think), you might want to refer them to Bob Dorsey's amusing photo series (http://www.davedorsey.com/dna.html) showing them exactly what to expect. He takes the viewer through the entire process (Figure 9-1) in comical detail that will quickly relax even the most wary.

Figure 9-1: Bob Dorsey takes his DNA test.

© *Kathy Peacock*

Guaranteed Results

An interesting phenomenon that most administrators will confront at some point is a prospect who expresses no concerns and seems willing to join the project, but wants assurances that they will receive immediately useful results. Usually, this means they want to be sure that they will match someone else in the study. Of course, there's no way you can promise this. Over time, as your project swells, the odds become greater that any random testee will have one or more instant matches. But as with all genealogy, there are no guarantees.

Setting expectations is essential to ensuring that participants will not be disappointed if their results make them an unrelated "orphan" within the project, even if only for a short time. When Janice McGough (Gordon) encounters people who will only test if they can be sure they will match someone, she explains, "If they are sure they should be matching someone, the cost is small compared with hours and money spent trying to find a connection that may not even exist. I try to stress to everyone that this is a young project, and we may not have a match right away, but with patience, everyone should be able to answer some genealogical question and further their research."

It also helps to emphasize the pioneering aspect of this kind of research. Georgia K. Bopp (Kinney and variations) uses an honesty-is-the-best policy approach and historical analogy to convey this point on her Web site:

"As for you, you are similar to one of the first people who signed up for the telephone. It's new. It's expensive. Sometimes it does not work right. When you talk to the phone company, they don't have answers to your questions—or they use so much jargon that you have no idea what they are talking about. And even on a good day, there's hardly anyone to call, much less anyone you really want to talk to. And not only that, most of your friends think you are a fool and wasting your money. On the other hand, you are ahead of your time, a real pioneer, an adventurer, on the cutting edge of a whole new technology. Hopefully, your pioneer spirit includes patience and a sense of humor!"

Our mini-industry is already migrating out of the pioneering stage, our projects are continuing to proliferate in size and number, and our recent recruits are increasingly savvy, so the need to place quite so much emphasis on expectation-setting is diminishing, but it's still wise to inform new participants of the no-guarantees aspect of this testing as a precaution.

Fear of Surprises

Some people love surprises, but others can't stand them. And with DNA testing, there's always the potential for unexpected revelations. In fact, 64 percent of project managers we surveyed have encountered at least one I-can't-believe-this situation, and quite a few have run into two or more such surprises. Most prospective participants won't even think about this unless they happen to be experienced genealogists, but it's in your best interest to inform them of this possibility before they sign on, so they're prepared for it should they be on the receiving end of some shocking news!

By far, the most common surprises stem from non-paternity events. Just because someone bears a given surname does not ensure that there hasn't been a "duped daddy" or unknown adoption in the line of descent sometime over the past few centuries. In such cases, the startled descendant may be stunned to find his DNA markers do not match anyone else with his surname, even though he's long been an active participant in the family association. A project manager may find himself having to explain to a pair of first cousins why their DNA doesn't match (yes, it happens). Or someone who has been researching their line for 35 years may be disgruntled to learn that he's not descended from Swanson X as his diligent research shows; he's really descended from Swanson Y from the neighboring county. Or someone whose ancestor had been rumored to be illegitimate or adopted may learn that that's not the case. And it's not just non-paternity events that can produce astonishing news. A person who considers himself Caucasian might learn, for instance, that he has a predominance of African-American DNA, or vice versa.

All of these are real-world situations, and if your project grows to any size, you'll probably bump into one or more of them yourself. Whether these are happy or unpleasant surprises depends on the perspective of the person being tested, but it's useful to heed the old "forewarned is forearmed" adage. Tell your participants that this chance exists up-front, so it's less shocking if it does occur. Include this fact in your FAQs or early communications. Better to lose a potential participant than to have him learn something he's not prepared to handle.

Fortunately, many participants revel in surprising news, or at least accept it without protest because they, like most genealogists, are after the truth. In fact, many administrators report that those who received out-of-the-blue results eventually admit that they had heard family stories and had some suspicions going into the test. That may even be their unvoiced reason for joining the study. Every once in a great while, a testee will resent the results and insist that your science, not their research, is wrong. Regardless of the circumstances, the best approach is to respond to whatever the reaction is with compassion and a willingness to listen and answer questions. If you're respectful and responsive, you may find that the shocked recipient ultimately becomes one of your project's biggest supporters once he's had a little time to assimilate the information.

Money, Money, Money

But what about money? Surely that must be a cause of resistance. Yes, you're right, although it's less of an issue than you might suspect. Still, because it is a frequent obstacle, and there are so many different ways to tackle it, we decided to consider it separately from other possible forms of resistance.

Regardless of how participants are recruited, one question will inevitably arise: Who pays? Prices for DNA testing have come down (at the time of this writing, costs ranged from $80 to $570 per test or battery of tests—mostly in the lower end of this spectrum), but are still considered a luxury by some. Genealogists are accustomed to spending money for their research—$15 for this death certificate, $19.95 for that book, perhaps even $69.95 for an annual online subscription—but some still get sticker shock at the cost of DNA testing. And if the individual concerned is not a genealogist, they may be even more perplexed about why they might want to spend this money

Experienced project managers point out that DNA testing can actually result in serious cost savings. Steven C. Perkins (Parkins/Perkins) explains to potential participants that "It is far more efficient to know your DNA ancestry and then concentrate your time, effort, and money on related families than to keep looking at everyone with your surname. It will save you money since you'll be able to spend only for the materials that deal with your genetic family."

And this is very true. Imagine how many years and dollars would have been spent trying to connect all the Smolenyaks who (as we learned back in Chapter 3) are not related, if a handful of tests had not proven otherwise. Indeed, this cost-saving potential is one of the least recognized benefits of DNA testing.

Who Pays?

In an ideal world, everyone would "get it," and there would be no need to be creative about financial considerations. But the reality is that many of us are careful with our money and may need a little encouragement to part with it. Or we may just not be willing to—period. Or we may be very young, very old, unemployed, on a fixed income, or otherwise unable to pay, even though we're interested. So what are your options? We turn once again to our panel of project managers for the approaches that have worked for them. If you're an administrator, you can try these suggestions:

- allow participants to self-select (via the broadcast strategy) and pay for their own tests
- proactively recruit participants and request that they pay for their own tests

- fund the study yourself if you have deep pockets (admittedly a rare occurrence)!
- seek a philanthropist within your DNA community to pay for some or all of the tests (yes, a few have succeeded in this)
- join forces with an existing family or surname association, and obtain funding earmarked for specific testing purposes
- ask most to pay, but personally subsidize or pay for them on a case by case basis (e.g., if they're the only living person with a DNA sample that's critical to your study)
- provide a scholarship program on your Web site to subsidize or pay for selected tests (ask a core group of interested people or your association to kick in perhaps $25 each for tests critical to the study or to outright pay for some)
- conduct a fundraiser (through your association or at a family reunion) with the purpose of collecting money just for testing
- ask for sponsors for individual tests (women who can't be tested for Y-DNA may pay for a relative)
- offer an installment plan to those who are keen to participate and willing to pay, but just can't pay all at once
- ask would-be participants whose DNA would be redundant in your study to "adopt a line" by paying for someone else who is willing to provide a sample, but unable to pay
- remind prospective participants that their DNA represents their extended family, and suggest that they get their siblings, cousins, and other relatives to contribute to the cause
- try to talk a university into studying your clan (works best if you happen to belong to a population they were already keen to research)

Gregg Bonner (Bonner and Lentz) is even more creative than most and is trying to establish a kind of brokerage service to match DNA samples to payers. As he explains, "I think there are a lot of people like me who would be willing to pony up money for certain samples that represent us by proxy. They believe in the tests and are willing to pay for some, but can't themselves just fund everyone. So I'm trying to develop a matching service where people can pledge $10 (let's say) for a particular sample, and when the pledge fund exceeds the costs, then the test proceeds."

James Reynolds Hull (Hull and Reynolds) uses a similar approach and has added a Web page (www.hullsurnamednastudy.com/page7.html) allowing interested parties to donate to testing via PayPal. You may also find, as he has, that people who are initially reluctant may pay after the fact:

"A lady whom I hoped would sponsor a family member told me she didn't believe in 'it.' After we had quite a discussion about Alexander Graham Bell, Thomas Alva Edison, and Henry Ford, and I was not too successful in convincing her that many people didn't believe in voice over wire, electricity, and the automobile, she immediately began complaining about the cost. When I told her that I was so convinced that DNA could prove or disprove whether her Hulls were related to the Crewkerne Hulls that I would personally pay the fees, she asked me why I would want to do that. I replied because her Hull ancestors want her to know one way or the other. She said, 'Well, okay, if you want to do it, how can I object?' The donor's results connected her Hulls to the Crewkerne Hulls. She was very happy and later sent me a check for the fees."

A Little Help from My Friends

We've covered the most frequent questions, concerns, and objections raised by potential participants, but should you encounter a situation not covered here or just want more insight, we suggest seeking help from your peers. A posting on Rootsweb's popular GENEALOGY-DNA Mailing List (http://lists.rootsweb.com/index/other/DNA/GENEALOGY-DNA.html) or a visit to www.dnalist.net, a resource designed to assist DNA project managers, will undoubtedly produce some excellent ideas from others who have walked in your shoes.

10

Interpreting and Sharing Results

You've taken the plunge, and now you have some results in your hands—an enigmatic string of numbers or letters. This chapter will cover ways to add more meaning to them. For the most part, this translates into comparing your results with others', and you can invest as little or as much effort as you'd like.

Many DNA tests are done for the simple purpose of determining if person A matches person B. In such cases, a quick glance at both people's results will answer the question, and—ta-da!—the analysis is complete. But many of us are curious to learn more. "Hmm," we think, "I match Fred. I wonder if we match anyone else." And so we set about seeing what else we might be able to learn. In this manner, it's fairly common for us to work our way through several levels of analysis. After comparing with selected individuals of particular interest, we take our collection of numbers or letters and look for them in these places:

1. our DNA project
2. any relevant testing company databases
3. public access databases and Web sites
4. published technical literature

It's not always in this order, but this is the most typical pattern. All through this book, we've been ignoring the "ladies first" adage by discussing Y-DNA before mtDNA, but in this chapter, we're going to reverse our usual sequence and start with mtDNA. We're doing this because mtDNA analysis is more straightforward and will allow us to briskly demonstrate this common pattern of stair-stepping our way to additional insight.

Because it is so often used as part of formal surname studies, Y-DNA analysis frequently involves an extra level of investigation. Somewhere in the midst of this research process, something we discover (perhaps a person with a similar surname who's only one mutation off from our haplotype) might prompt us to go a step further and do a little number-crunching or data consolidation in the hope of extracting still more meaning from our initial results. And for better or worse, there are more ways one can consider, inspect, and tinker with Y-DNA results, so we'll linger on the Y chromosome for a while in order to explore a few of the more popular tools and calculations. (In case you're curious, we're referring to genetic distance, MRCA, and the like.)

While it's only an approximation, if we were to expand upon the basic four-step research process above and try to capture a more complete range of analysis possibilities, it might look something like Figure 10-1.

Different people have different styles. We share this not to suggest that everyone should march their way through every box. Many people—in fact, probably the majority of people involved in genetealogy—are content to remain in the upper left-hand box. We should also mention that this is not intended to be comprehensive. A thorough consideration of all possible techniques, coupled with examples, would fill a book itself. Rather, the intent is to provide a brief survey of options. Should you be

Figure 10-1: Analysis options

Level of Analysis			
	Basic	**Intermediate**	**Advanced**
Within Study/Testing Company	Test results/Web site for your study—seek perfect and close matches only and compare pedigrees; review ready-made reports (e.g., genetic distance)	Test results/Web site for your study—cluster all participants; combine with genealogical data; MRCA calculations	Determine ancestral haplotype; display results in network diagrams or genetic distance grids
Beyond Study/ Testing Company	N/A	Public access databases/Web sites: Yhrd, Ybase, Ysearch, SMGF mtDNAlog, mtDNA concordance DNAPrintlog	Research technical literature for information pertaining to your haplotype, haplo-group, etc. (e.g., Google, PubMed)

one of those who decides to venture beyond basic analysis, we hope you'll refer back to this chapter from time to time for more ideas. Consider it a menu of possibilities. You can satisfy yourself with an appetizer or go for a full five-course meal!

Finally, we'll conclude this chapter with a discussion of tactics for sharing your results. As we mentioned earlier, effectively communicating results has multiple benefits. It increases participants' understanding and enthusiasm, attracts new participants, and improves your chances of solving your particular mysteries and learning about your ancestral past. And of course, the more we all share, the more we all learn!

Mitochondrial DNA Analysis

What can you do with your mtDNA data after you've framed your certificate and memorized your differences from the

Cambridge Reference Sequence? If you were testing to try to solve a genealogical conundrum, your first (and if you like, last) step is simply seeing if you have a match with your testing partners. If you have a match, you have fresh and compelling evidence in support of the theory you developed. If not, it's back to the drawing board to formulate a new theory to test.

But let's say you match, but want to know more, or you've done an mtDNA test just out of curiosity. What can you learn from your testing company? Some of them will report if you have matches with other customers—your genetic cousins—but random matches are unlikely to forge new genealogical links. This is particularly true if you have one of the more common haplotypes. The penalty for the "success" of your haplotype is an excess of distant cousins. Because the mtDNA molecule can remain stable over thousands of years, the Most Recent Common Ancestor (MRCA) could have lived several thousand years ago.

However, it doesn't cost much to exchange data with your mtDNA cousins. It's like purchasing an inexpensive sweepstakes ticket—your chances of winning are very low, but the entertainment value is high, and once in a blue moon, there's a winner in the "random match lottery." Richard Ferguson shared his startling experience: "I had a complete match on the mtDNA test. I thought it would be another one where the match was too far back, but when I checked, we both had a female ancestor named Smith. That still sounded doubtful until we compared the families and discovered that they lived in the same county

When swapping data with your maternal cousins, don't get hung up on surnames, since they change with marriage every generation. Instead, focus on commonalities in location and time frame. If you both have ancestors who lived at the same time in the same vicinity, your chances of a closer connection ratchet up, even without a surname in common.

in Alabama and shared an unusual first name, Didama. Ultimately, we determined that his Smith way back there was the mother of my Smith, so the DNA test took me another generation back on my maternal line."

Not all testing companies maintain databases, and most databases tend to be limited to customers who tested with them. Companies are also bound by privacy agreements, so contact information is excluded unless the customer has given explicit consent. Family Tree DNA reports exact matches to customers (although there is no way to browse the database), and Oxford Ancestors allows its customers to search for any haplotype as well as exact matches. The resulting display lists the name and e-mail address (if provided), the current residence, the deepest-known maternal ancestry, and the "clan" (haplogroup). Oxford Ancestors also has a public access database with limited features (www.oxfordancestors.com/ms_guest.htm). You can select a clan name from a drop-down menu and get a list of records with current residence, deepest-known maternal ancestry, and number of variations from the Cambridge Reference Sequence. Oxford Ancestors' message board (www.oxan.click2.co.uk/cgi-bin/index.php) has forums dedicated to each haplogroup as well.

mtDNA Databases

Person-to-person connections through mtDNA may be hard to find, but background information is not. If you decide to go beyond the information your testing company furnishes, you have several options. Your thoughts may turn first to Google—always a good choice—but two other resources are the mtDNA Test Results Log Book and the Mitochondrial DNA Concordance.

Unfortunately, a full-featured public access database for mtDNA—one that allows searching records on multiple fields—has yet to be designed. In the meantime, Charles Kerchner has opened a guest book (www.mtDNAlog.org) where people can deposit their results. The log is structured with fields for name, date, testing company, HVR1 and HVR2 results, haplogroup, date of test, contact information, and free-form comments. The comments may include ancestral names and locations or special interests, such as Jewish or Native-American studies. You can skim the log (very lengthy, but fun to read) or search for keywords with the *Find* function of your Internet browser. A good way to search for HVR mutations is to *Find* the number for one of them (e.g., 16223).

The Mitochondrial DNA Concordance is another, especially helpful resource. A concordance is a reference tool, an index that displays an alphabetical list of keywords in their full context, typically in works of literature such as the Bible or Shakespeare. The line "What's in a name? that which we call a rose by any other name would smell as sweet" will be displayed multiple times, under "name," "rose," "smell," and "sweet."

The Mitochondrial DNA Concordance (www.bioanth.cam.ac.uk/mtDNA/) is a collection of several thousand mtDNA haplotypes collected from scientific literature, up until 1998. The file is broken up into a number of Web pages, each covering a certain range of HVR1 or HVR2. (The HVR2 section is very limited, however, and not coordinated with HVR1 data.) Exact matches to the Cambridge Reference Sequence are very common, found all over Europe, so they are not included in the concordance.

The concordance lists polymorphisms as "words" in the context of a full haplotype. For example, if your haplotype is 16293[G] 16311[C], you can start your search in the section of the table that lists either the word *16293* or the word *16311*. Both will present you the same listing. Because the Web pages are very large, you'll probably want to use your browser's *Find* function for *16293[G] 16311[C]*.

A fragment of the table listing for the keyword *16293[G]* is shown in Figure 10-2. Since mtDNA changes so slowly, only exact matches are relevant, with a couple of exceptions.

For instance, the older research sequenced a more limited range, often from 16024 to 16365. If you have differences both inside and outside of this range—say, at 16293 and 16519—you might still want to consider any 16293-only listings from older

Figure 10-2: Extract from Mitochondrial DNA Concordance

Polymorphisms	Author; year; sample ID; origin
16024-16089[0] 16293[G]	Richards, 96; 9; Portuguese(1)
16024-16089[0] 16293[G] 16311[C]	Richards, 96; 396; Cornish(1)
16293[G]	Bertranpetit, 95; 16; Basque(1) Sajantila, 95; sovtoi; Finnish(1) Còrte-Real, 96; 46; [1]; Portuguese(1)
16293[G] 16311[C]	Horai, 90; SB32; Cauc. UK(1) Miller, 96; NOR.0025; Norwegian(1) Miller, 96; NOR.0029; Norwegian(1)

articles. Also, the notation 16024-16089[o] means that those bases were omitted from the research article.

The table shows that the haplotype 16293[G] 16311[C] has been found in Cornish, English, and Norwegian samples, providing a hint of your possible maternal origins. Of course, your ancestors may never have lived in any of these places since the database shows the location of people living today, but it's not unreasonable to assume that at least some of your mitochondrial ancestor's descendants stayed put.

Given that the database contains several thousand records all told, this haplotype is apparently quite rare. If more samples were acquired, the haplotype could show up in other locations, but from what's presented here, it seems to be centered in Northern Europe. In contrast, the 16293[G] haplotype is found in the Iberian peninsula as well as Finland.

PubMed

Perhaps you've already consulted your testing partners, testing company, the mtDNA Test Results Log Book, and the Mitochondrial DNA Concordance, but *still* want to know more. Then you're ready to move on to the technical literature. PubMed (www.ncbi.nlm.nih.gov/entrez/query.fcgi?db=PubMed) is a massive index to the technical literature at the National Library of Medicine—and although we're focusing on mtDNA for the moment, this resource is also useful for Y-DNA.

At this site, you can search for articles with information about your haplogroup or background. Try searching on *mitochondrial dna haplogroup t* or *mitochondrial dna jewish, or mitochondrial dna gypsy* (or Africa, Native American, etc.), and you'll see a list of citations. If a title strikes your fancy, click on the hyperlink to bring up an abstract. The abstract page has several useful links too. One is "Related Articles," which will often bring up articles you didn't find with the particular keywords you chose.

Another link is called just that: "Links." One menu choice will list articles that cited the one you're looking at—a good way to locate more recent developments. Another menu choice is "Books." Clicking on that will reformat the abstract, with hyperlinks on technical terms. Clicking on the term will then bring up a list of pages in online textbooks for more detailed explanations. The abstract page will often have a link to the journal publisher as well, useful because some publishers place the full text of journal articles online several months after publication. Alternatively, you can Google the words in the title to see if the author has uploaded a copy to his personal Web space.

Don't be put off by the technical level of the articles. It's not necessary to understand every single term to harvest fascinating tidbits from them. The discussion section of an article often summarizes findings and concepts in plain English. Besides, you can just look at the pictures if you wish! Charts and diagrams are worth more than a thousand technical words.

We've just walked through the basic analysis process as it pertains to mtDNA, but this isn't the last you've heard of mtDNA. Later when we discuss sharing options, we'll tell you about an innovative approach that's being used specifically for mitochondrial results, but we turn next to the more plentiful options for Y-DNA analysis.

Y Chromosome Analysis

You've waited several weeks, and the moment has arrived! You've received your test results and are staring at a bunch of numbers that looks something like 13–24–16–10–14–15–11–13–13–13–11–32–17–8–10–11–11–26–15–20–32–14–14–15–15. Now what?

The first layer of analysis is the easiest, yet perhaps the most important. This cluster of numbers is your haplotype, so now you're equipped to play the haplotype-matching game. If you had a particular testing partner, you'll start by comparing results with him. Your mystery will finally be solved (Yes, you do share a common ancestor!), or your mismatch will let you know that there's a flaw in your traditional research or testing hypothesis. But most of us will start by comparing haplotypes with all the others in a surname study in which we're participating. Do we match anyone? Have we found some genetic cousins?

And if you are using a testing company that maintains a database of all its Y-DNA customers, you may also be informed of other exact or close matches you have beyond your own surname project. For instance, Family Tree DNA customers who sign a release can obtain names and e-mail addresses for possible genetic cousins who are also open to exchanging information. This may be a useful avenue to pursue, particularly if your haplotype is relatively rare.

Many people are content to stop at this point, but others want to wring the last drop of information out of that string of numbers. If you're a drop-wringer, there are plenty of ideas in this chapter especially for you, starting with public access databases.

Y-DNA Databases

The next layer of analysis is just an extension of what you've already done. First, you compared results with your fellow project members and then with others in your testing company's database. Now you're just going to poke around in other databases scattered around the Internet. For Y-DNA, you have several options, and it's worth taking a look at all of them:

- www.yhrd.org
- www.ybase.org
- www.ysearch.org
- www.smgf.org

WWW.YHRD.ORG

The Y-chromosome Haplotype Reference Database (Yhrd) was developed for forensic use, but genealogists have a way of sniffing out novel resources. It is an anonymous database, with random samples collected from different locations. "Forensic" in this context doesn't mean the samples came from criminals, but the database was designed for courtroom testimony about the frequency of haplotypes. It originated in Germany, so many of the records are concentrated in that country. However, the scope is expanding, and most European countries are represented as well as some African and Asian countries and ethnic groups in the United States (Caucasian, Hispanic American, and African American).

For an interesting exercise, try entering the results from the controversial Jefferson-Hemings case, published in the technical journal *Nature* in 1998. Much debate continues about the interpretation of the results, but here we'll limit ourselves to the raw data, displayed in Figure 10-3.

Clicking on the *Search database* link at www.yhrd.org brings up several ways to search the database. Select the *GeoSearch* option, then the *Haplotype* option. You'll see a data entry screen with drop-down boxes for various STR markers. Plug in the values for descendants of Jefferson's brother-in-law, Dabney Carr, then click on Search. It will bring up over 400 matches and display a may of locations where the halotype was found. They are scattered all over Europe, from sunny Spain to the fjords of Norway. Clearly this is a common and widespread haplotype.

How about Jefferson? One of the key points in the 1998 article was the rarity of Jefferson's haplotype, which had not been found in a reference sample of 680 Europeans. The Yhrd data-

Figure 10-3: Raw data from Jefferson study

DYS	19	389-1	389-2	390	391	392	393
Carr	14	13	29	24	10	13	13
Jefferson	15	12	27	24	10	15	13

base now contains more than 20,000 records. Does the Jefferson haplotype still stand alone? Yes, it does. This is not unusual—even a database with 24,000 records has not exhausted the great diversity of haplotypes.

If you have a haplotype that's in between, not too common and not too rare, it may show up in just a few geographic locations. As with the Mitochondrial DNA Concordance, these locations should be taken as an indication only.

WWW.YBASE.ORG

Ybase was the first open-access database designed especially for genealogists. Sponsored by DNA Heritage, it can accommodate results from any genealogical testing company. One can search for a surname of interest or a haplotype. Entering Carr's haplotype brings up several hundred records. Many of these are clusters from one surname study, but the surnames range from Arnold to Young. This illustrates the limitations of haplotypes based on a small number of markers—you end up with a surplus of random matches.

Ybase also has some statistical summaries about the frequency of alleles on individual markers. If you have a rare haplotype, as Jefferson did, you can gain some insight into just which markers make you distinctive.

WWW.YSEARCH.ORG

Ysearch is sponsored by Family Tree DNA but open to all. Existing customers can add the results from their personal results Web page with a single click, but anyone can search the database and add records manually. The data entry form will accept results for markers used by any of the testing companies. It also has fields for some biographical information about the most distant known ancestor.

The database is searchable by name and haplogroup. You'll get hundreds of hits if you ask for Haplogroup R1b (the most common one in Europe), but if you have a rare haplogroup, it could be interesting to compare notes about geographic origins with your clan cousins. After selecting a record from the name or haplogroup list, you can then *Search for Genetic Matches*, demanding strict matches or allowing partial ones. A tab for *Research Tools* will generate a grid of Y results for records you select and calculate genetic distance for you.

WWW.SMGF.ORG

One of the newest databases has a broader mission than just the Y chromosome. You may not be aware of the Molecular Genealogy Research Project (MGRP) of the Sorenson Molecular Genealogy Foundation (SMGF)—or you may be among the more than 40,000 people who have already donated a DNA sample to the study. The project's goal is to collect more than 100,000 samples from individuals who can provide ancestral charts (with complete birth dates and places) dating back to at least the 1800s. DNA submissions are analyzed for Y, mtDNA, and an impressive 300 autosomal markers, and the plan is to ultimately correlate all this genetic and genealogical data.

Fortunately for all of us, the project intends to share its findings and took the first major step in 2004 with the launch of its Y-DNA database. You can read more in the next chapter about exciting developments to expect from SMGF in the future (and how to participate in the study, if you'd like), but for now, we'll focus on the Y database.

It was built using the Y-DNA of the project's more than 12,000 male participants to date. The initial release included about 5,000 men with data on 24 selected markers. Additional markers will likely be added if research indicates that they are genealogically relevant.

In the interest of privacy, no personal identifiers are included. The only details associated with the Y-DNA of donors are names, places, and birth dates of their ancestors born prior to 1900—and if you're fortunate enough to have a match, that may be a gold mine for you! The database can be queried by simply entering your own Y-DNA markers. Any matches are returned in the form of genealogical data pertaining to those pre-1900 individuals in the database sharing the specified markers.

With this information, you can then use traditional research techniques to try to ascertain any connection to your family (although the linkage could predate the paper trail), or perhaps to locate and join forces with others researching these ancestors. Alternatively, you may find a strong geographic pointer, indicating where you should focus future research efforts. Initially, those with rare haplotypes may find no matches, but as the database continues to build, that should change, especially given that SMGF is making a concerted effort to collect DNA samples from more than 500 populations around the world. Surely your population will be represented at some point!

Beyond Databases

As with mtDNA, you may wish to explore still further by Googling or searching PubMed for relevant articles. To get a feel for what you might find, try searching on *y chromosome jewish, y chromosome Africa,* or *y chromosome STR.*

And as we mentioned earlier, when it comes to Y-DNA, there are more ways to play with your results than with mtDNA—more tools and calculations you can use to mine your data for every possible ounce of meaning. Much of the bulk of this chapter will be devoted to this topic, but you should be aware that if you want this level of analysis, it's not necessary for you to do it all yourself. Some testing companies will do it for you. If you're not a do-it-yourself kind of person and anticipate a rather sizable project, you might want to consider a consultant.

For instance, Relative Genetics' approach is to work closely with a project coordinator, who supplies genealogical data to correlate with genetic data. The company then prepares a final report (in PDF format) that includes the original pedigree, the actual pedigree as determined by DNA results (maybe the same and maybe different!), a network diagram showing connections between haplotypes, and several other customized charts. There are several Family Genetics™ success stories listed on their Web site. For an example of a lengthy report, have a look at the Barton surname project (www.bartondna.info/Report_2003/BartonDNA.pdf), which ended up with 13 lineages for 107 participants.

Other companies meet you halfway by providing some ready-made analysis for you. Family Tree DNA, for example, provides a list of locations where similar haplotypes have been traced as well as a genetic distance report from the perspective of each and every participant in your study.

Genetic Distance

Sooner or later, it happens to all of us. We encounter a close, but not perfect match for our haplotype—perhaps within our project or our testing company's database. And once it happens, you're introduced to the concept of genetic distance.

The phrase "genetic distance" can be used in several contexts. Sometimes it's about kinship—second cousins are more distantly related than first cousins. A molecular geneticist calculates the physical distance between genes on a chromosome, with units in millions of bases. A population geneticist looks at allele frequencies of multiple genes in various populations and computes a score for the overall similarity of different groups, based on those percentages.

Family Tree DNA also computes a genetic distance score as a way of reducing the differences between two haplotypes down to a single number that's easier for all of us to grasp. The genetic distance is an estimate of the number of mutations required to get from one haplotype to another. Since mutations are rare events, the number of mutations is an indirect way of estimating the degree of kinship. Closely related people will tend to resemble each other more than distantly related people. A group of 100 pairs who are brothers will exhibit fewer mutations than a group of 100 pairs who are 10th cousins.

As you might suspect, a genetic distance of zero means that the two haplotypes are identical, but let's look at a common situation where a pair of men have close, but not identical results. In Figure 10-4, the genetic distance is two. The direction of the difference—whether a marker has gained or lost a repeat—is not important. The genetic distance is the sum of the absolute values (so both +1 and -1 are treated the same). In this case, we simply add 1 and 1 and arrive at the genetic distance of 2.

Direct observation of many father/son pairs has shown that most mutations are a change of one repeat. Figure 10-4 shows a straightforward example, but there are a few quirks to genetic distance calculation. You can confirm your answer by entering the two haplotype profiles at www.ysearch.org, then clicking on Research Tools and Genetic Distance Report.

Most Recent Common Ancestor

When two people match—when they have a genetic distance of zero—they realize that an ancestor's name is not spelled out

Figure 10-4: Genetic distance = 2

DYS Marker	**393**	**390**	**19**	**391**	**385a**	**385b**	**426**	**388**	**439**	**389-1**	**392**	**389-2**
Haplotype #1	14	22	14	10	13	14	11	14	11	12	11	27
Haplotype #2	14	22	13	10	13	14	11	14	12	12	11	27
Difference	0	0	-1	0	0	0	0	0	+1	0	0	0

in the DNA alphabet, but they may still harbor a hope that their DNA results can tell them *when* their Most Recent Common Ancestor lived. In fact, this is usually also true when two individuals have a close match with perhaps one or two mutations separating them. They know that their family trees overlap at some point in time, so it's only natural to wonder when.

Fortunately, it's possible to make an assessment—not in precise terms, but within a range. Mutations are not only rare, but random, and that randomness takes us into the fuzzy world of statistical probabilities. Statistics are good for describing general tendencies in large samples, but they can be challenging to apply to individual cases.

So why trouble with MRCA calculations? It's a matter of prioritization. Using MRCA input, you can make smarter decisions about whether it's worth pursuing a missing paper trail after you've seen the DNA results for two people. Three mutations in 10 markers results in numbers that indicate your time and effort might be better spent elsewhere, but one mutation in 30 to 40 markers narrows the time frame considerably, suggesting that you might have a reasonable chance of finding the answers you seek in the paper trail.

MRCA Calculation Factors

The units of measurement for MRCA calculations are mutation rate, number of markers, and number of generations. Another factor that plays a role is prior knowledge. For instance, knowing that those being tested have the same surname is a plus when it comes to MRCA calculations.

Mutation Rates

Mutation rates have been studied in hundreds of father/son pairs, in deep-rooting pedigrees with a few persons representing many generations, in shallow but broad pedigrees with many descendants of a few men, and even in thousands of sperm collected from a few men. The range of values is from zero (none detected in the sample) to 0.004, but the overall average when all data are pooled hovers around 0.002 (0.2 percent) per marker per generation. Another way of stating this mutation rate is one change in 500 generations for each marker.

Number of Markers

The various testing companies offer a range of markers, a factor which will likely influence your choice of companies. More markers are always welcome, but there's a cost-benefit trade-off to consider. The Y chromosome doesn't have enough STR markers to tell us *exactly* when a common ancestor lived, but matching on more markers gives us greater confidence that the Most *Recent* Common Ancestor was indeed recent. Naturally, tests with more markers cost more, so what is the optimum number to test to ensure meaningful results?

In the final analysis, it comes down to the specifics of your situation. For example, in a study sponsored by one of the authors, the first two candidates were tested on 25 markers. Their shared haplotype proved to be exceedingly rare, even considering a limited subset of those markers, so a 12-marker test was deemed sufficient for a third candidate. Rare haplotypes need fewer markers to give one confidence in a match. By contrast, someone with a common haplotype and many matches at the 25-marker level may wish to upgrade to still more in order to narrow the field of genetic cousins to those who are more closely related.

A well-designed project focusing on one ancestor often searches for widely separated candidates to test. If the budget is limited, it might be preferable to test three candidates on fewer markers than two candidates on more markers, in order to confirm the original or ancestral haplotype (a concept we'll explore more shortly). On the other hand, a broad study may need more markers to rule out false-positive matches. This would be especially true of a study such as the Welsh patronymics project (www.small-stuff.com/WELSH), where many participants are likely to have a haplotype common in Wales. In general, if money is no object, the more markers, the merrier! But if you're like most of us, the particulars of your project will steer your decisions about how many markers to have tested.

Generation Length

We're not used to generations as a unit of measurement. Instead, we record a date of birth for an ancestor and perhaps note that he lived 250 years ago. If we want to convert generations to years, we're adding another layer of uncertainty to the MRCA calculations. The sons on the pedigree chart in Figure 2-1 were born when their fathers were 29 and 22 years old. If you check your own pedigree chart, you'll undoubtedly find an even wider range, because men can sire children while in their teens and even into their later years.

Anthropologists and population geneticists do not agree on a standard number of years for generation length, although 20 and 25 years are often mentioned. They are dealing with different cultures and time periods, both factors that affect the average. Marc Tremblay at the University of Quebec studied generation lengths in a large collection of French-Canadian parish records and genealogies extending back as many as 13 generations. The average age at the birth of the first child was about 24 years, and the average age at the birth of the last child was about 46 years, so he concluded that a more realistic number for the past few hundred years would be about 35 years per generation. This seems consistent with the personal experience of many genealogists, but you may hail from an ethnic or social group that tended to marry later or have smaller families. It's helpful to simply be aware of this variable because many of us like to convert generations to the most MRCA into years.

Prior Knowledge

Since MRCA calculations assume randomness, the results can sometimes seem a little disheartening. Left to our own devices, we would probably almost always underestimate the time back to the MRCA, so we may not like those unrepentant numbers that tell us the range of possibility is greater than we'd like to think. But the rules of the game change if you can add some prior knowledge to the mix. In genetealogy, prior knowledge usually means a shared surname or evidence of a common geographical origin. In short, if you have an exact or close match with someone with the same surname or who has documented roots in the same small village as you, it's more meaningful than if you discover a match with a random person from, say, one of the publicly available databases.

MRCA Calculation

One of us developed a utility program that allows you to enter any mutation rate and any number of markers. The program then estimates the median number of generations to the Most Recent Common Ancestor for two people. That is, 50 percent of matched pairs would find their common ancestor within that time frame, but 50 percent would have to keep on searching.

If you'd like to experiment with this MRCA calculator, you can download it for free (members.aol.com/dnacousins/MRCA.exe), but for your convenience, we've summarized results for 10 to 50 markers in Figure 10-5.

After all the cautions about the fuzzy nature of statistics, you may be amused at the precision implied by the decimal points in Figure 10-5—not to mention the concept of fractional generations! Nonetheless, there are important insights that are quite apparent from this table. Adding more markers clearly reduces the number of generations to the MRCA. The number of generations drops substantially going from 10 to 20 (an indication of why so many people upgraded to tests of more than 20 markers when they became available), but the decline is less dramatic for each new set of 10 markers added. And allowing for one or two mutations markedly broadens the distance to the MRCA.

Ancestral Haplotype: Bridging the Mutation Gap

There's another circumstance that can hint at a closer relationship than the genetic distance might indicate by itself. Figure 10-5 shows only a 50-50 chance of encountering the MRCA within 35 generations if there's a genetic distance of 2 on a 20-marker test. Yet many surname projects have numerous instances of participants in the same boat, even when the common ancestor is *known* (that is, documented) to be much more recent.

This puzzling paradox is often resolved when the *ancestral haplotype* is determined. The two participants may differ from each other by two mutations, but each one differs from the common ancestor by only one, as shown in Figure 10-6. The ancestral haplotype is a bridge between the two participants.

The ancestral haplotype is easy to deduce when two or more sons have descendants with identical haplotypes. If you look at Figure 10-7, for instance, you'll see that E's haplotype is evident if I and J match, but B's haplotype is not proven (since there could have been a mutation or even a non-paternity event). A match between I and K will confirm B's haplotype (since B is their closest common ancestor), and a match between I and M will lead you all the way back to A's haplotype.

It's more of a challenge if there's no consensus haplotype in the different branches, but it's still possible. In such cases, the next step is to look for a consensus value on individual markers.

Figure 10-5: Median number of generations to MRCA

Number of Mutations	Number of Markers 10	20	30	40	50
0	17.3	8.7	5.8	4.3	3.5
1	44.3	21.5	14.2	10.6	8.5
2	74.8	35.2	23.1	17.1	13.6

Assuming average mutation rate = 0.002

Figure 10-6: Bridge or ancestral haplotype

Haplotype #1	11	11	11	12	11
Bridge/Ancestral Haplotype	11	11	11	11	11
Haplotype #2	11	12	11	11	11

Figure 10-8 contains data from the Ausburn/Osborne study (freepages.genealogy.rootsweb.com/~tlosborne/AusburnSurname Project). All markers were identical except for the ones shown here, DYS449 and DYS464. Thomas is descended from one son, Lecil and George from two branches off a second son, and Maynard from a third son. All four men had a different haplotype, but taking the value that appeared most often for each marker (frequently referred to as the majority rule) gives a haplotype of 29–15–16–17–17. That happens to match exactly with George's haplotype, which lends support to the existence of such a haplotype. (That's not essential, but one must be cautious about this majority rule approach. If there are many participants from one branch compared with others, they may unduly influence the majority.)

Thomas and Lecil differ by two, but they each differ by only one from the bridge haplotype of their common ancestor. In this

Figure 10-7: Deducing ancestral haplotype

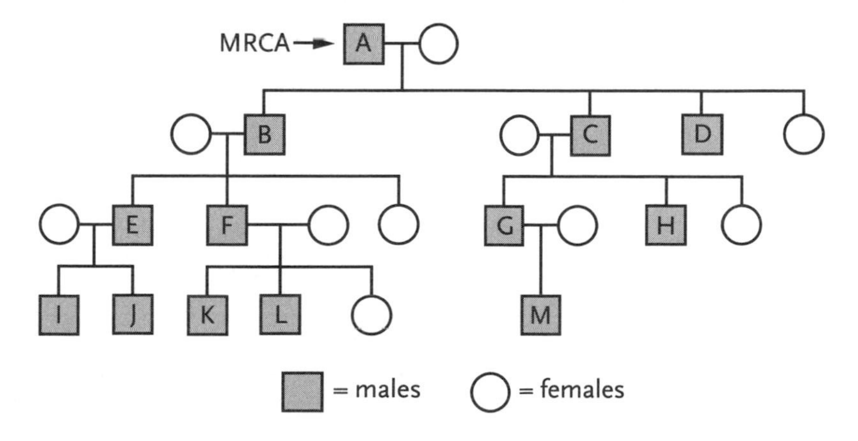

example, the genealogical connections are known from the documentary evidence, but if there is no paper trail, it may be worth looking for a new donor to build the bridge. If the parties with two mismatches can locate distant cousins to test, they may be lucky enough to find a DNA donor to fill the gap.

Mutations in various branches are often welcomed as a diagnostic sign. If someone new joins a surname project without knowing his ancestry, and he has a mutation that matches a branch off the ancestral haplotype, he can immediately place himself more precisely in the descendancy chart.

But Ausburns will need to be cautious if new participants looking for the right branch have mutations that occur in the DYS464 system. DYS464 is a *multi-copy* marker—it occurs at several different positions on the Y chromosome. It's not possible to tell which allele is found at a particular position on the chromosome, so the results are simply given in ascending order.

The ancestral haplotype has two markers with a value of 17. Lecil only has one 17, so either 464c or 464d mutated. No matter which one mutated, Lecil's results would be presented in ascending order 15–16–16–17. A new person with the 15–16–16–17 haplotype has no way of knowing whether he truly matches Lecil or if his haplotype is the result of a mutation in the other 17.

Figure 10-8: Alleles for four descendants

			DYS		
Name	**449**	**464a**	**464b**	**464c**	**464d**
Thomas	28	15	16	17	17
Lecil	29	15	16	16	17
George	29	15	16	17	17
Maynard	29	15	16	17	18

More Tools, More Toys!

If you're the kind who likes to really play with his data, there are many other tools for analyzing and displaying it, such as these:

- network diagrams, which visually show connections between people as points and lines (See fluxus-engineering.com/sharenet.htm and www.roperld.com/RoperGenetics.htm#network25 to learn more.)
- a mutation calculator, which shows the probability of encountering mutations in a study group (Visit members.aol.com/dnafiler/MutationCalculator.exe for a free utility program written by one of the authors.)
- genetic distance grids, which allow study participants to see their closest matches at a glance, much like charts included with maps showing the distances between a variety of cities. The Ausburn study shows examples where the spelling of Ausburn/Osborne did not predict the closest matches! Dean McGee has shared his methods for creating a genetic distance grid using Excel (www.mymcgee.com/tools/index.html).

You can learn more about these and other techniques and toys on the GENEALOGY-DNA mailing list (lists.rootsweb.com/index/other/DNA/GENEALOGY-DNA.html). In fact, the mailing list is like having hundreds of consultants at your beck and call, ready to render an opinion on your case. It's the Swiss army knife in your toolbox.

Sharing Results

Enough of all this analysis! Genetealogy is no fun and almost pointless in a vacuum, so now it's time to share your results. How can you do this? The most basic approach is simply communicating results with your project members, but you can also contribute data to public access databases, or as many project administrators have done, launch your own Web site.

Communicating

If you're managing a project, it goes without saying that you'll want to share results with your participants. You'll likely find yourself exchanging private e-mails with many of your members, but at some point, that may become rather cumbersome—particularly if you are fortunate enough to have a fair-size study under way. For this reason, many administrators opt for a blend of personal e-mails, public postings on relevant surname message boards, and mailing lists (e.g., at www.rootsweb.com, www.ancestry.com, www.genforum.com, etc.), and even newsletters (electronic and/or printed copy). In fact, some who rely primarily on a Web site periodically broadcast by these other means to inform others of Web site updates and gently remind them to stop by for a visit.

How broadly should you define the audience for these communications? As broadly as possible—without disturbing anyone who's previously indicated that they're simply not interested. Clearly, you'll want to provide updates to those who have actually been tested, but including those you *wish* would get tested is a useful way of keeping them apprised of the project's progress. And after a few success stories, they just might be tempted to join.

Everyone is busy these days, so keep your messages, postings, or newsletters relatively short—and be sure to write in plain English! Don't expect recipients to be excited about the discovery of an ancestral haplotype if you don't tell them what it is and why they should care. And if you include actual results, take the time to explain all those numbers. It helps to think back to how you felt when you first gazed upon your initial results. Use your experience to shorten your readers' learning curve by simplifying whenever possible and using examples. Make liberal use of color-coding, bolding, and other techniques for emphasizing important data so they don't have to strain their eyes to figure out what matters. Most of all, help them understand what the data you're sharing means to them. Genetealogy can seem like a foreign language to some at first, and you are their translator.

Adding to Databases

Remember earlier in this chapter when we told you about database resources you could search for matches? Luckily for us, some of them—www.ysearch.org and www.ybase.org, in particular—are two-way streets. Not only can you search the data; you can also add to it. So if you want to spread the word, sharing your results in these databases is a logical step to take. As the popularity of genetealogy spreads, these databases are apt to become even more important in terms of attracting fresh participants to your study because they're the first places many newcomers will check to see if someone is already researching their surname.

But what if you want to share your mtDNA results? There's the mtDNA Test Results Log Book we mentioned earlier, of course, so you might want to start by adding an entry there. But wouldn't it be nice if there was an easy way for all of us to upload our mtDNA data in conjunction with our genealogical data, so we could all find out if any of our matrilineal relatives have already had mtDNA tests? Gregg Bonner thought so and decided to do something about it. He's proposed a mechanism for associating mtDNA profiles with genealogical databases at WorldConnect (see http://freepages.genealogy.rootsweb.com/~gbonner/mtDNA), which offers free hosting and indexing services for personal genealogy databases. Gregg writes about his vision:

> *"Many people have had their mtDNA sequenced in an effort to better understand their ancestry. Generally, the reports associate the mtDNA sequence only to a particular person (i.e., the mtDNA sample provider), or at best to the matriline from the sample provider to the matriarch (the sample provider's mother's mother's mother's (etc.) mother, as far back as is known).*
>
> *"But this type of reporting underutilizes the technique. Clearly the mtDNA descent is known from the matriarch down to the sample provider, but its descent is also known*

> *along* all *the lines of its descent. So once an mtDNA type is known for one member of an 'mtDNA' family, that mtDNA type is known for* all *the members of that mtDNA family. So there should be some way to compile a list of names of people who should have a particular* ***known*** *mtDNA type. With a collection of these types of compilations, it is hoped that people can search the files and find their matriarch or a possible matriarch . . . Of course, this will only work if we have enough datasets to look through to make it useful. And since nobody is doing this now, I have decided that I will do it."*

Gregg's Web site has instructions on how to upload your mtDNA family to the WorldConnect Project. For a sample of how this might work in practice, pretend that one of your ancestors is Iris Brockway. If you look for her name at WorldConnect (worldconnect.rootsweb.com), she will pop up in a file called bonnie-mtdna (note the clue in the file name!) Clicking on that entry will bring up the individual record for Iris in Bonnie Schrack's compilation of matrilineal relatives. The header on the page begins, "Everyone in this file should have mtDNA type I1: 16129A, 16172C, 16223T, 16311C, 16391A, 16519C."

If you then click on *Pedigree,* the maternal line of Iris is displayed. Follow the maternal line as far as it goes, then click on "Descendancy." Everyone listed on that page has the same mtDNA. You'll note that, like other participants, Bonnie generously included material she collected on many collateral lines. In fact, she's even expanded the scope of her interest to general background on Haplogroup I (www.ancientrootsresearch.com). Such Web sites dedicated to a haplogroup are yet another way to share.

Web Sites

The broadest outreach, available 24/7, comes from publishing on the Internet. If someone Googles your surname and DNA, they will find you. To design a Web site that will not only share results but also help attract new participants, take some time to browse what others have done, and borrow some ideas from the sites you like best. Although there are many excellent DNA project Web sites, a particularly outstanding example is the Blair DNA Project (blairgenealogy.com/dna) managed by John A. Blair. Cleanly designed, it includes these features

- Home page: to introduce the project and orient the reader
- DNA 101: an overview on Y-chromosome testing that's so well executed that many other project managers provide a link to it
- Test results: succinctly captured and color-coded project results to date
- Ancestors: basic life details about the earliest known ancestors of the various testing participants (allowing a new viewer to see if perhaps his or her ancestor is already represented in the study)
- Participants: contact information (as much as they're comfortable sharing) and links to pedigrees for test participants
- Blair DNA FAQ: answers to the most commonly asked questions about DNA testing, some specific to the Blair project but many more general in nature
- Status report: a timeline summarizing milestones in the project
- Application: the application form to join the study and be tested
- Pedigree chart: a form in which a participant can enter his pedigree so that they're all provided in a uniform format
- Info release form: a form allowing the participant to decide which information (name, pedigree, e-mail, address, phone) he is willing to publicly share

- Release of liability: a release protecting the project manager against any claims of harm
- Marker analysis: an overview of what different levels of matching (e.g., 12 of 12, 11 of 12, 25 of 25, etc.) mean in terms of generations to the MRCA
- Blair Society lines: a summary of results for Blair Society for Genealogical Research lines represented in the DNA study and others for which participants are sought

You may also wish to have a look at the following projects to find other ideas to use for your own Web site:

- Mumma Surname DNA Project (www.mumma.org/DNA.htm): the pioneer surname project from which many have borrowed analysis and reporting techniques
- Graves Surname DNA Study (www.gravesfa.org/dna.html): an extensive project with impressively summarized results and success stories
- Hatcher Surname DNA Study (http://homepages.rootsweb.com/~nhatcher/DNAlinks.htm): a project that collects records on stray Hatchers along with DNA results
- Kerchner Surname Y-DNA Project (www.kerchner.com/kerchdna.htm): another well-organized site with a particularly useful overview (click on *Genetics 101 and Intro Report*) of DNA testing
- Lindsay Surname Project (www.clanlindsay.com/dna_project.htm): a well-analyzed project containing an especially comprehensive background on many aspects of DNA

If you were to closely inspect the content of the surname project Web sites that appeal to you most, you'd almost certainly notice two features in all of them. The first is a clustering of participants by haplotype (often, but not always, listed under the name of the earliest known ancestor) so that members can easily grasp the connections the DNA research has revealed or confirmed. Most sites accomplish this through tables where identical and close results are presented together and color-coded. Mutations may be bolded or otherwise highlighted so the viewer can quickly spot differences. Dean McGee's Web site (www.mymcgee.com/tools/index.html) shows how to automate this process with Excel macros.

A second valuable feature is pedigree information for testing participants, and it's fast becoming standard practice to link directly to this information from the haplotype tables. Incorporating this genealogical data allows participants to compare notes and look for connections with their haplotype mates, makes it possible for others to determine if perhaps their Y-DNA is already represented in the study, and enables the project manager to somewhat reduce his middleman role and make better decisions about whose participation to actively seek.

The few Web sites we've mentioned are just the tip of the iceberg. An easy way to visit many sites is to start at Pieter Cramwinckel's webring (j.webring.com/hub?ring=dnasurname projec). Some have been developed by people who clearly enjoy the process of Web design for its own sake, but don't let that intimidate you. Family Tree DNA provides a template for creating a basic Web site in just a few minutes. If you want to elaborate on that, you can begin in a simple and straightforward way by pasting a table of results into a Word file, adding a few comments, and exporting it in HTML format. If you build it, they will come!

Now that you've analyzed your results to the point of exhaustion and disseminated them to every corner of the globe, you may be wondering "What's next?" Funny you should ask. That's the subject of the next chapter.

Part 4

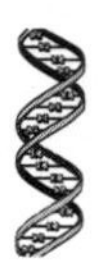

The Future

11

What's Next?

We should probably begin this chapter by confessing our bias. In case it isn't extremely apparent yet, we are pro-genetealogy. And when we conducted research to help us peer into the future, we contacted a mixture of more than 150 DNA project managers, testing company principals, scientists, and professional genealogists. Not surprisingly, the majority of these people share our bias, so our collective forecasting may strike you as a little on the optimistic side.

Having said that, we're comfortable with the predictions we're about to make because of what we said at the outset about the tipping point. Genetealogy started at a crawl, steadily gathered momentum, and in our estimation, is about to reach that stage where it spreads like wildfire. And once you arrive at that point in terms of any new technology, it's virtually impossible to go back. We just don't see how this genie could be shoved back in the bottle. So yes, we're bullish on genetic genealogy, but not without reason.

Growing Popularity of Genetic Genealogy

Technology guru Dick Eastman has more foresight than most. He wrote the following in an article that appeared in *Ancestry Daily News* on January 3, 2001, back when it probably sounded borderline absurd to some:

> *"As we gain more knowledge about genetics and as more genetic 'maps' are created, and also as more powerful computers and online databases become available, I predict that genetic profiling will become commonplace. If all of our relatives, including very distant cousins, have DNA samples made and recorded, future computers and databases will be able to map our relationships in detail. By matching the millions of bits of information together, we should be able to someday establish a person's ancestry with great precision. Someday we will accept genetic maps in lieu of pedigree charts."*

As he readily acknowledges, he doesn't expect this to happen overnight, but at least within the lifetime of most of those reading these words—and it seems we're well on our way.

There's been an initial reticence with genetealogy—more so than with most technologies. Compounding the usual fear of mastering a new tool is the web of misconceptions about DNA. "Provide a sample of my DNA?" many think, "Why don't I just put my credit card details on the Internet while I'm at it?"

But as curiosity gets the better of some, the success stories pile up, and people educate themselves about genetealogy, we come to realize that we *can* master the technology and that furnishing a DNA sample won't come back to haunt us. As DNA literacy grows, it will become common knowledge that genealogical testing is not the same as that used for criminal or medical purposes (and we hope now that you've read this book, you'll help spread the word!). We will also undoubtedly develop more sophisticated techniques for protecting our genetic privacy should the need arise, but for the present, measures such as posting results under the names of ancestors who lived more than a century ago are working just fine.

The barriers to participation are evaporating, and as they do, genetic genealogy is exploding. Bennett Greenspan, owner of Family Tree DNA, predicts that "DNA testing will be integrated into all family genealogies within 10 to 15 years, and within this decade, all English speakers will have at least one DNA project underway or completed." DNA project manager Wade Glascock (Hooper) asserts that "DNA testing will soon be as common a genealogical tool as checking the census or going to the courthouse." Molecular genealogist Diahan Southard of Relative Genetics echoes this opinion, saying, "We are only at the very beginning of genetic genealogy. I see it becoming as much a part of genealogy as any other historical document."

In fact, genetealogy is very much following in the footsteps of its older brother, genealogy, which grew beyond all expectations over the past decade, largely due to the Internet—and all indications are that the two should play well together. DNA has already proven to be a valuable tool for genealogists, especially when used to support or strengthen evidence from available records. At other times, it steers stumped genealogists in the right direction. As Alastair Greenshields, one of the founders of DNA Heritage, likes to explain, "Genealogy is a bit like playing Clue. The facts are usually all there to be found, but when you get stuck, DNA whispers in your ear and says, 'Look in the billiard room!'"

Project administrator Mary Lou Hudson (Cox, Anderson) also points out that having this younger sibling around will help his older brother rectify some of the excesses of his rapid growth: "The Internet is one of the best things that's happened for genealogists, but it's also created a monster that only DNA testing can resolve. Family trees are passed around and posted to the Internet without verification. The quickest and easiest way to verify that a tree is correct will be through DNA testing."

Ugo A. Perego, Senior Project Administrator with the Sorenson Molecular Genealogy Foundation (SMGF), underscores this reliability factor: "Because of the inheritance properties of DNA, we are literally a walking, breathing, living record of our family history. You can change, destroy, or hide records proving relationships between individuals, but you cannot change the genetic makeup we have inherited from our biological ancestors."

Those who pursue their past are on a quest for the truth, and genetealogy provides a powerful tool to confirm the research we've already done, to give us a hint when we're stuck, and to make sure we haven't been taken in by some misinformation. For all these reasons, we are confident that we will be welcoming countless genetealogists into the fold in the very near future.

Preserving Samples

"If only I had asked her when I had the chance." How many times have we heard this phrase or one like it? It probably won't be long before you hear the latest addition to this repertoire of regret: "If only I had snagged a sample of his DNA when he was alive." Just as people realize that they've lost a living history book when an elderly relative dies, they will come to understand that a piece of their genetic history can vanish just as easily. Maybe Great-uncle Joe was the last in his line, and his Y-DNA—representing your grandmother's line—is now beyond reach. If it happens to you once, you're not apt to let it happen again.

Consequently, one likely side effect of the growing popularity of genetic genealogy will be a corresponding growth in sales of DNA preservation kits as a form of insurance. Even though testing prices are coming down, some consider them too steep or just can't afford to test as many people as they'd like to for their projects. An easy and inexpensive preventive measure is to obtain kits—as little as $25—to collect samples now for later testing. Many kits require just a simple swab of the cheek and use a preservative (FTA™) for easy storage at room temperature. There are no excuses! You may want to consider requesting samples from elderly relatives all at once and then gradually submitting them for testing over time.

Expanding Scope of Projects

As we become more comfortable with the ins and outs of genetic genealogy, it's not surprising that the scope of many of our projects is growing. We may start with a modest goal—say, to see if we share a common ancestor with someone else of the same surname—but once we get that question answered, we tend to move on to the next. Perhaps we decide to open our project for anyone with our surname. Over time, that project slowly gathers steam, until it finally reaches critical mass—the point, as DNA project manager Roy Hutchinson (Hutchinson) describes it—"when there are enough family lines established that the non-genealogist can take a test and have a good chance of determining which line he belongs to." At that stage, a surname project can become largely self-sustaining in the sense that the benefits of participation are more obvious, so the administrator no longer has to aggressively recruit newcomers.

This is a typical evolution for surname projects, especially for ones focused on fairly common names. But increasingly, our studies are expanding in other respects. We may start with Y-chromosome testing, and then venture into mtDNA or Bio-Geographical testing. Our single surname study may blossom into a multi-surname one in order to learn more about those who lived in the same village or region as our ancestors. Or perhaps we may decide to construct a genetic pedigree, obtaining Y-DNA and mtDNA samples to represent each of the branches of our own family tree back for a selected number of generations. (See www.calabriadna.com/Louis-tree.html for Louis Loccisano's easy-to-follow genetic pedigree.) Maybe we'll even do all of the above, as one of us already has.

Whether the expansion occurs through more participants, more tests, more surnames, more objectives, or more branches of our family tree, the key word is *more*. Just like genealogy, genetealogy is a slippery slope. We may intend to dabble, but we get pulled in. There's a beneficial, self-nourishing cycle at work here. The more we learn, the more we want to know. The more we want to know, the more tests we take or sponsor. The more tests we take, the more results we have to share with others. The more results we share with others, the more we learn from our collective pool of data. And the more we learn (you guessed it), we're back to step one and launching ourselves through the cycle again. Add to this the numbers that will be pulled in through the public's growing awareness of genetealogy, and you can sense how inevitable the acceleration and intensification of this cycle is. Within a few years, DNA testing to learn about one's ancestry will be routine. Need something for Mom's birthday? Why not a DNA test? You can't get much more personal than that, and it's the gift that keeps on giving—even to future generations!

Faster and Cheaper

Really? Yes, really! In the not-too-distant future, the testing process will be quicker, and prices will be lower. Much of this is due to the usual economic factors—greater demand translates into more competition, which leads to lower prices and other enhancements. In fact, anyone who's been involved in genetic genealogy has already seen this principle in action, but first-timers (and even experienced project managers) are often frustrated by the time it takes to get results.

In our I-want-it-yesterday world, having to wait for a few weeks to a few months doesn't sit too well with many of us. Fortunately, speedier turnaround time is one of the improvements we can look forward to. Terry Carmichael of Sorenson Genomics (dbaGeneTree and Relative Genetics) plainly states that "Results will be released much more quickly than they are today. There is a lot of improvement that can be done to make this type of DNA testing more efficient. In the past, much of this testing has been offered by academic institutions using low-throughput equipment. Today, high-throughput and accredited laboratories . . . are working on developing methods to do the testing more quickly, efficiently, and economically."

Cost can be looked at from two angles: the actual price one pays or what one gets in return for his money. While we expect prices to continue to decrease as they have over the past few years, we imagine a point at which they will plateau, but we'll get more bang for the buck—that is, more markers will be examined, so the cost per marker will come down. The same amount that used to buy you X markers will buy you X + 5 markers, and those extra markers will provide you more meaningful results. You'll see the same charge on your credit card statement, but you'll gain a little more insight than you would have if you had taken the test several months earlier.

Bennett Greenspan believes that, "Like computer prices,

Easier Recording and Reporting

As genetealogy grows in popularity, additional support tools will naturally appear. For instance, while it's the exception at the moment, it will rapidly become standard practice for genealogical software packages to include fields for DNA data. Before long, they'll also include the ability to generate charts specifically designed for genetealogical purposes. Looking for a potential mtDNA candidate to represent an ancestor who lived 200 years ago? Simply highlight that ancestor, and select the "mtDNA descendancy chart" option to see which of your distant cousins are eligible. In the meantime, T. Mark James has written a freeware GEDCOM viewer that will display everyone in the file who is related in the Y or mtDNA lines, in the form of a collapsible tree (http://freepages.genealogy.rootsweb.com/~tmark/GedView.html).

We also expect that it will become the norm for testing companies to provide Web site templates to make it easier for their customers to share results with project members, as Family Tree DNA does. Gradually, these offerings will become more sophisticated, providing tools to facilitate more advanced analysis, such as Most Recent Common Ancestor calculators (which will take into account mutation rates of specific markers and do the math for all members of a project) and phylogenetic networks as described in Chapter 10.

DNA testing prices will come down and features will go up." But he cautions that, "The challenge is to ensure that the pricing of this type of test doesn't get so low that companies can no longer afford to help their customers understand what the results mean." And that's a legitimate point. If we as consumers want the testing companies to keep raising the bar in terms of the services they provide, we can't expect them to offer $10 tests. The usual cost-benefit tradeoff exists, and we would argue that support services in genetic genealogy are more important than for many other products. Lower prices, yes—but not so low that no one answers your next e-mail or phone call!

What DNA Project Managers Want

When we surveyed DNA project administrators, we were curious to see the future they envisioned, the forward steps they anticipated. Sure, we all want everything faster, cheaper, and easier, but what else? What emerged was essentially a project manager's DNA testing wish list, which we've summarized below:

- What I look forward to in the very near future is established mutation rates at individual markers, which will improve calibration of the mutation clock and result in a better choice of markers.

- I am hoping that we need to retest all lines with 50-marker tests with enough mutations that will sort out the relations that aren't clear with 0 to 2 mutations on fewer markers.

- I hope to see Y-DNA and mtDNA tests that will be detailed and specific enough to not only link you to your kin, but to explain the relationship to other kin. And to have a hierarchical key to enable you to tie your Y and mtDNA results down by country and even by lineage.

- I hope the science will progress so that a test will become available for cross-gender testing. Right now, when a male line is interrupted by one female, the Y-chromosome test cannot be used.

Based on input we gathered from genetic experts, the good news is that most of this will come to pass. In fact, we presented this list in what we perceive to be descending order of probability. The higher up the list, the more likely and the sooner it will happen. Let's take a closer look at some of these predictions.

More and Better Markers

Most project managers are somewhat marker-obsessed. If you look at our online chit-chat, perhaps half of our discussions pertain to markers. How many are enough? What constitutes a close match? How quickly do they mutate? And so on. Underlying all of this is a hunger for as many ancestrally useful markers as possible.

Those who have been involved for several years are familiar with the upward trajectory of markers. Initially, we were thrilled with 10 or 12 markers. But after a while, we learned that the results from such tests could occasionally be misleading. And while there's still a place for such low-resolution tests (especially when cost is a major concern), we expect that they will be phased out. Alastair Greenshields shares the concerns of many who have come to question tests with fewer than 20 markers: "They can provide far too many false-positives and false-negatives—that is, wrong conclusions. If people don't progress to high-resolution tests, then there is the real possibility that false genealogies will be created, 'supported' by DNA evidence. This is especially important in surname projects because most people with the same or similar names originate from a particular country or region, and the potential to have similar haplotypes when only a dozen markers are analyzed is greatly increased."

But it's not just the number of markers that are examined that matters; the selection of markers also plays an important role. The markers we currently use were obviously not picked at random. They were chosen based on desirable characteristics, so it's to be expected that the majority of those used by various testing companies overlap. But after a few years of layering documented genealogies on top of DNA results, it's become clear that not even all our chosen markers were created equal. Diahan Southard, for one, is confident that we'll be able to squeeze more out of the markers we currently use: "There is much work to be done in refining the current process and using the information more effectively. For example, there is so much more to a 26-marker Y test than just if there are enough matches to warrant a relationship. More research will be conducted to determine specific mutation rates and allele frequencies in order to more accurately pinpoint the exact relationship between two individuals, whereas right now, the MRCA calculation is more of a calculated guess."

Ugo Perego notes that "There are several studies currently underway that are evaluating the need of specific markers or the use of additional markers for Y chromosome testing. It appears that some markers currently offered in commercial Y testing are not polymorphic within a specific population, and therefore, the information they provide doesn't add much to the whole picture. On the other hand, it is possible that some markers not yet standardized and used in Y chromosome testing might provide insights about groups of people not detectable using the current suite of markers available commercially."

Alastair Greenshields agrees that in general, "the more polymorphic, the better," and believes that research into marker mutation rates will be beneficial in this regard. He continues, "I'd also imagine that multi-copy markers would be valued less and less for genealogy. The very nature of this type of marker means that the laboratory performing the tests cannot know which repeat number belongs to which marker, so the possibility of forming a false conclusion is greater."

So in the coming months and years, we'll benefit from both quantitative and qualitative progress. And moving in parallel with the growing sophistication of the tests themselves will be our ability to extract more meaning from the results. "At between 30 and 50 Y-STR markers, the best testing panels will have sufficient resolution for most genealogy situations, except adoption," says Bennett Greenspan. "With well-established mutation rates, family genealogists will be able to work with tighter

DNA Extraction

It's already happened. There's no one—not even a fifth cousin once removed—who has the "right" DNA for your study. Grandpa was the last in your family who did, and he passed away 15 years ago. But wait! You saw that show on television where they got DNA from a stamp that had been licked, and you have some old letters from Grandpa. Could you get his DNA from one of those envelopes?

Maybe, but you will likely experience a little sticker shock, and even if you're lucky, you'll probably only get mtDNA. Newer technology allows for the extraction and analysis of smaller DNA samples than what we were capable of in the past. However, there are other issues in addition to the size of the DNA sample left behind by an ancestor, the main one being contamination, where an item containing or carrying the DNA of a dead relative is handled improperly, and the DNA becomes contaminated by those handling it.

But don't get discouraged. Experts generally agree that advances in science will almost eliminate the need to test deceased individuals or items associated with them. And if you still really want to get that old cap Grandpa wore every day tested, there are companies that will do that for you (see Appendix B) and companies that will accept DNA samples obtained in this manner.

time frames, higher confidence levels, or both." It's hard to say how fast and far we'll ultimately go, but considering genetealogy would have qualified as science fiction less than a decade ago, it's remarkable that we've already reached a point that more than satisfies many people's curiosity about their ancestral past.

What about mtDNA and SNPs?

Alastair Greenshields has a few predictions and words of caution about the future of mtDNA and SNP testing: "A breakthrough for mtDNA will be made when the entire circular strand of mtDNA is routinely tested in many diverse populations. This will clarify the picture already built up with what we do know about mtDNA. This is still quite expensive to do. The sections of mtDNA that are studied usually involve the hypervariable regions, which as the name suggests, vary a lot. Another advantage to these regions is that they are also non-coding (i.e., don't code for proteins). Having data from the whole mtDNA strand will increase the understanding of the origins and movements of maternal lines, but great care must be taken when selling such detailed tests to the public because many of these new regions are medically informative."

He continues, "SNP tests, which are of huge interest to population geneticists, will also make a greater impact on genealogy. Whenever people receive their Y-STR results, they not only want to compare their STR haplotypes with others, but they also want to glean as much information from their results as possible, which includes guessing at their haplogroup. The advent of several techniques for multiplexing SNPs allows for easier determination of a person's haplogroup. Although SNP tests can't produce the finer resolution pictures that STR tests can, the strata of haplogroups and history of human migration are fascinating topics to pretty much everyone who has their own DNA tested."

Terry Carmichael also sees a bright future for both mtDNA and SNP testing: "In the next decade, we will be able to match mtDNA sequences and Y-SNPs to a vast number of different populations, such as Asian, African, Northern European, Southern European, and even more accurately to their subpopulations. Databases of DNA populations will also assist us with answering questions regarding global human geographic origins, paternal and maternal migration patterns, historical environmental conditions, and more."

Our take? We look forward to significant improvements in both of these areas, especially with mtDNA with its 16,569 base pairs. In scientific terms, this is a very manageable number, so we expect that one or more DNA testing companies will offer a complete mtDNA sequence product. The question is timing.

Will someone offer it to us when the cost to conduct the testing (and consequently, the price) is still high? Or will they wait until it can be done in a more cost-efficient manner, so the price can be lower and resulting demand would be higher? And a not so incidental secondary issue is the medical aspect. Since this would likely be the first genealogical product that could potentially reveal truly confidential information, the privacy aspect will require greater attention.

As to SNPs, it's hard to imagine why there wouldn't be considerable progress in identifying more and affiliating them with various populations, but it will take time. Richard Villems, Director of the Estonian Biocentre at the University of Tartu, explains the challenges involved: "As far as Y chromosome is concerned—given its length and molecular evolutionary speed, it can be predicted that even a father and son will differ in about one to two single point mutations (SNPs). This means that a true genealogical tree of mankind can be reconstructed at a very deep resolution—theoretically at least—but technologies that are already available cost far out of any reasonable scale. In a more realistic scale, DNA chips with about 5,000 to 10,000 NRY SNPs can already be constructed now, but an extensive, costly, and complicated organization and understanding of the pattern of the tree, covering the world at a reasonable frequency, is a huge undertaking. Many, if not most of the globally interesting SNPs might well not have been found yet, and many others are just 'private'—not really helpful for genealogy or for population genetics." For now, testing companies will undoubtedly be on the lookout, and the most avid of genetealogists will continue to snoop through scientific papers for the latest SNPs that might be of interest for our purposes, so progress for us on this front is apt to come more as a trickle or minor stream than as the big wave we might prefer.

Enter Autosomal Testing

During the first couple of years of genetealogy, Y-chromosome and mtDNA testing were our only options (except for those of us with special situations we wished to address through paternity, avuncular, or other such close kin testing). And while Y and mtDNA alone can provide fascinating insights into our roots, they can only reveal information about the two outermost lines of our pedigree chart, or less than 1 percent of a person's total DNA. Ugo Perego likens using just these two to "looking into a keyhole and trying to describe the beautiful room on the other side of the door in great detail."

Inspecting selected portions of our autosomal DNA is not quite the equivalent of opening that door, but it's at least tantamount to cutting a huge window in it. The possibility now exists to learn so much more than before, and the first genealogical test to offer this opportunity was DNAPrint Genomics' ANCESTRYbyDNA product.

As with its purely Y and mtDNA counterparts, this test is becoming increasingly sophisticated. First, it examined about 70 markers to enable us to learn our genetic admixture at a macro level. Findings were broken out by percentages of Sub-Saharan African, Indo-European, East Asian, and Native American. Then it was upgraded to 170 markers to provide greater accuracy and precision in these results. The next forward step will be another version that will allow us to learn our results at a finer level of ethnicity as well as get a better understanding of whether our admixture was of recent or ancient origin. Your 100 percent Indo-European, for instance, might be reported as 40 percent Mediterranean, 30 percent Eastern European, and 30 percent Scandinavian, and your Native American might be broken down into tribal affiliations.

As scientists continue to identify markers that are particularly good at distinguishing assorted populations, we will undoubtedly benefit from ongoing improvements to such geographic origin tests. Tony Frudakis, Ph.D., of DNAPrint Genomics expects that other companies will follow in their footsteps and that even more specific information will be available in the future: "It's not that the answers you have now are written in stone and tell you everything you want to know. More support tools will come to bear." By way of example, he adds, "We could type a thousand Norwegians, a thousand Italians, a thousand Russians. We could do this for hundreds of populations, and you could then go through all these profiles to see which one fits you."

Of course, costs and potential demand for such tests will drive whether they actually come to market, but genetealogy will certainly profit from its ability to piggyback on research and data developed for other purposes. For instance, scientists are keen to explore how susceptible people with differing ancestries are to certain diseases and how responsive they are to various drugs. This important biomedical application naturally creates a need to be able to delve into people's ancestry—information that's a gold mine to us. Companies such as DNAPrint Genomics didn't develop their tests as genealogical products, but they recognized this potential use and fortunately made the de-

DNA Adoption Banking

Linda Hammer has made it her life's work to reunite people and devotes much of her effort to the millions whose lives have been touched by adoption. She is launching the DNA Banking Project. Adoptees and birth parents are encouraged to participate by providing a cheek swab sample and signing a release. Once the bank includes enough samples, the matching process—and ultimately the reunions—will begin.

Linda says that the power of the bank is that adoptees and birth parents will be able to find one another without so much as a name, date, or place of birth. This aspect is particularly important for those who were adopted from overseas or through black market adoption rings. She cites, for example, the fall of Saigon during which 3,000 babies were airlifted out and shipped to Australia, France, Canada, and the United States. Those wishing to learn more can visit www.the-seeker.com/dna.htm.

cision to put it into the hands of the everyday person. With a bit of luck, we will continue to be the accidental beneficiaries of such spin-off offerings!

But beyond this, what might be possible in terms of autosomal DNA? You probably won't be surprised to hear that researchers disagree. Currently, there are no tests available to trace ancestry across all the branches of your family tree, aside from the fairly broad ANCESTRYbyDNA test we just discussed. Many are not optimistic about using autosomal DNA for family history, but the Sorenson Molecular Genealogy Foundation believes it is possible. According to a spokesman for the foundation, "Scientists are studying the possibility of using SNPs, STRs, or a combination of the two. We believe this can be done and are working toward that goal. These studies will open a new era in the field of genetics and genealogy." We touched on their Y-surname database in the last chapter, but what exactly is the SMGF and its Molecular Genealogy Research Project (MGRP)? In their own words, the main purpose of the MGRP "is to create a database that will allow people to enhance and extend their genealogical research through the use of genetics. Not only are we looking at the Y chromosome and the mitochondrial DNA inheritance patterns, the MGRP's database will help trace family lines following the inheritance pattern of autosomal DNA (the DNA from the remaining chromosomes)."

The project is aiming to collect both genetic and genealogical data—DNA samples and pedigrees—from 100,000 volunteers and is roughly at the halfway point. (We'll tell you how to participate in the next section.) The foundation is working on genotyping every individual at a rather astonishing 300 markers because it's thought that this number of markers "should be sufficient to connect people through their common ancestors."

Perhaps the most complex part of the process will be correlating all the data. As Ugo Perego explains, this entails "the development of complex algorithms that will perform statistical analysis and reveal how individuals are connected based on the number of markers they share because they were inherited from common ancestors." He elaborates, "The general idea is to be able to tell how closely or distantly any two individuals are related to each other. However, the first goal will be to link them to ancestors using autosomal DNA. It will be several years before this will be available. It's a very complex and ambitious approach, but we are confident that this can be done and that the results will be extremely valuable to the genealogical community."

While we're relieved not to be among those wrestling with the development of these algorithms, we like to think that the project will ultimately succeed, because it would represent a giant leap forward in our ability to learn about our ancestry through DNA testing. How would you as an individual access this mountain of data to peek into your past? By using SMGF's database, which leads us to our next area of speculation.

Age of the Database

Those of us who were among the first to get DNA tested often faced a frustrating situation once we received our results. We had a pile of numbers, and sometimes they confirmed or disproved the hypothesis upon which our project was based. But being curious types, we wanted to know more. We wanted to be able to compare our results with the big, wide world to see what extra nuggets of meaning we could extract.

Initially, our only option was to follow up on any matches indicated by our testing company's proprietary database, but since these were somewhat sparsely populated in the early days, the revelations were few and far between. Some even took to getting tested by more than one company just to have access to more than one database. Before long, we discovered www.yhrd.org, which we discussed in the last chapter. This was a step forward, and we were (and still are) grateful to be able to access this database designed and intended for the scientific community. But with limited markers and somewhat random populations, it can be a little hit and miss for our purposes.

What we wanted was a database for ourselves—one that was preferably open to all, not restricted to customers of one company or another. In this way, we would all be able to benefit from the largest pool of data possible. Before long, our wish was granted in the form of www.ybase.org. Alastair Greenshields, who announced this site in late 2002, remarks, "We created Ybase solely with genealogists in mind, and it contains both Y-chromosomal data and personal details of an ancestor. It can accept data from any of the current testing labs and continues to grow." DNA Heritage is the host of the site and continues to allow free access to all.

Also in 2003 came the introduction of two sites managed by Charles Kerchner where anyone could log their DNAPrint results (www.dnaprintlog.org) and mtDNA results (www.mtdnalog.org). These sites also allowed people to post extra details such as which testing company was used and how the results compared to their expectations. Since they're not databases per se, they require a little more effort to use, but they helped fill a gap since the few databases that did exist focused only on Y-DNA.

Fast-forward to late 2003, and Family Tree DNA's Ysearch (www.ysearch.org) made its appearance. It, too, is open to all. Family Tree DNA customers just have an easier time of uploading their data since much of the process is automated for them. Luckily for us, the company also opted to introduce extra functionality, such as the ability to search by haplogroup (the first time that STRs weren't getting all the attention!) and research tools, including the handy genetic distance report. We also appreciated the ability to add the name and birth and death dates of the farthest back known ancestor as well as information about immigrant ancestors.

And then along comes SMGF's Y/surname database, providing our first sample from the project's massive data collection effort and, it turns out, only the first stage in this database. Inaugurated with the Y-DNA data of 5,000 unidentified male participants, the database has been steadily growing, and the intent is to add not only new participants but additional markers over time. And since only people who could provide several generations of their family tree could join the study, names, places, and dates of birth of paternal ancestors born prior to 1900 (such information is in the public domain) are presented.

Ugo Perego provides an overview of how this database will evolve: "Our long-term goal is to have a database that will have Y, mtDNA, and about 300 autosomal markers for every sample collected. We would like these markers to be linked to the donors' genealogies so more complex ancestral and family relationships (beyond Y alone) could be studied. Currently, we have identified approximately 150 polymorphic markers that we think are good genealogical and geographic indicators. Some of these markers have been shown to be very useful within just a few generations, while others are more anthropological in use and can reveal information about someone's ancestry back many centuries. We are more concerned with markers that can be used to trace genealogical relationship within the past eight to 12 generations. We would like to link known genealogical information for ancestors born before 1900 to these autosomal markers, just like we are doing for the paternal Y." He cautions, however, that autosomal queries for genealogical purposes will not be available anytime soon.

As mentioned earlier, many are not convinced that autosomal DNA will work for genetic genealogy, but we fall into the optimistic camp. We are certain to learn many things of interest, even if the grand scheme remains incomplete. And even if it takes years to achieve, we already have much more than we could have envisioned at this stage. First we had nothing to eat, and now it's a feast every day! We have several powerful—and let's not forget, free—databases that continue to grow on a daily basis and provide us more and more insight with each tweak and upgrade. Will there be more databases? Possibly, but we don't need to be greedy. What we have right now coupled with the projected improvements is pretty remarkable. Will access continue to be free? Since there are several such databases, the answer is probably yes. Anyone who takes their database private is apt to lose visitors since there's more than one option. Will one database prove to be the most useful and essentially win out over the others? It's possible, but since they have complemen-

Scientists and Genealogists Working Together

If you are ever offered an opportunity to support a scientific study with your genealogical research in any way, please accept! The scientific community has been very open and receptive to genealogists because we have data that they find helpful. As Bruce Walsh of the University of Arizona summarizes, "Genetic analysis is greatly improved by having deeper pedigrees." And the reverse is also true. We're both in a position to help each other, and it's a convenient synergy.

Bennett Greenspan comments on a prime example, saying, "Scientific researchers truly need data for mutation rate estimates. That type of data will help scientists help us unravel time to our Most Recent Common Ancestor in an unambiguous fashion." In fact, Family Tree DNA merits particular kudos for its efforts to involve family historians in virtually all their innovations. Whenever they launch a new database, report, or feature, you can be sure that their customers were among the beta testers—and moreover, that their input and suggestions were taken seriously and implemented.

If you find yourself moved by an impulse to contribute to the cause and haven't received any invitations lately, one action you can always take is to participate in SMGF's study. While you used to have to attend a special event and give a blood sample, you can now do it by mail and just by swishing mouthwash (no more nasty needles!). Simply visit www.smgf.org, which will tell you how to order a free sample kit and donate a copy of your pedigree chart in support of this project. You won't get any information back for submitting the sample, but you'll get the satisfaction of helping fellow genetealogists, and you (or anyone else who cares to) will be able to query the database with results received from a commercial testing company. If you're feeling particularly inspired, order a few kits, and test other family members, because SMGF is seeking family clusters to participate. The site also explains how you can organize an event to gather still more samples.

tary features and data, there should be ample demand to provide plenty of traffic for at least a couple of them.

All in all, we're in territory that we could have only dreamt of even several months ago. The very existence of these databases is evidence that we've reached the tipping point in terms of genetic genealogy, and the possibilities they provide for genetealogical "newbies" should only serve to increase our numbers because they make it easier than ever before to look into our past.

No one gets DNA tested for the sake of getting tested. We do so to learn as much as we can about our roots, and while our "behavior" might still strike some as odd today, it will soon be commonplace. French philosopher Simone Weil once said, "To be rooted is perhaps the most important and least recognized need of the human soul." We agree that this underlying longing exists in every person, whether they can articulate it or not. How fortunate we are to live at the first time in the history of mankind when we have the ability to know the previously unknowable about ourselves and our forebears.

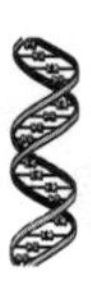

Appendices

Appendix A: Genealogical Resources

Societies

National Genealogical Society (NGS): www.ngsgenealogy.org

Federation of Genealogical Societies (FGS): www.fgs.org
P.O. Box 200940, Austin, TX 78720-0940
(888) FGS-1500 (347-1500); fax: (888) 380-0500

D'Addezio.com's Society Hill Directory: www.daddezio.com/society
Links to more than 5,000 societies.

Magazines

Ancestry Magazine: www.ancestry.com (click on "Shops" and enter "subscription"); (800) ANCESTRY (262-3787)

Family Chronicle: familychronicle.com; (416) 491-3699, ext. 111

Family Tree Magazine: www.familytreemagazine.com; (800) 448-0915

Heritage Quest: www.heritagequestmagazine.com; (866) 783-7899

Major genealogical societies also have their own publications, often both a journal or quarterly and a newsletter.

Books

While many bookstores have a strong selection of family history publications, here are several organizations that specialize in books and other products specifically geared for the genealogist:

Willow Bend Books:
www.willowbendbooks.com

Appleton's Books & Genealogy:
www.appletons.com

Heritage Creations:
www.heritagecreations.com

Genealogical Publishing Company:
www.genealogical.com

***Family Tree Magazine*:**
www.familytreemagazine.com/store

Ancestry.com:
www.ancestry.com (click on "Shop")

Forms

Most genealogical software packages incorporate the ability to automatically generate all of the standard forms (including the ones discussed in Chapter 2), so you may not find it necessary to investigate these online resources. We include them, though, because you might want to snag some forms before you select your software. And even if you've already made the purchase, a few of these sites have specially designed forms not available elsewhere. Invest a few minutes browsing, and you'll probably be rewarded with a couple of handy organizing tools that will move your research forward faster.

Ancestors:
www.pbs.org/kbyu/ancestors/charts
Includes pedigree chart, family group sheets, research log, source notes, and more

Ancestry.com:
www.ancestry.com/save/charts/ancchart.htm
Includes pedigree chart, family group sheets, research calendar, census forms, correspondence record, source summary, and more

Family History Library:
www.familysearch.org/Eng/Library/Education/frameset_education.asp (click on "Family History Library Publications" and then "Forms")
Provides an impressive menu of all the standard forms and many specialized ones

Family Tree Magazine:
www.familytreemagazine.com/forms/download.html
Furnishes both text and .pdf versions of popular forms, such as pedigree charts, family group sheets, research calendars, and correspondence logs, but also harder-to-find items including biographical outline, statewide marriage index, deed (grantee and grantor), research repository checklist, cemetery transcription, and artifacts and heirloom owners

Disney's *The Tigger Movie Create Your Own Family Tree* for kids (three forms):
www.disney.go.com/disneyvideos/animatedfilms/tiggermovie/familytree.html
Great place to start if you're young or just young at heart!

Software

If you're after a bargain, how does free sound? Both of the following can be downloaded for free from the Internet:

Personal Ancestral File (PAF):
www.familysearch.org
Click on "Order/Download Products" and "Software Downloads—Free." Then select the most recent version of PAF.

Ancestry Family Tree:
aft.ancestry.com
Offered by Ancestry.com, this software synchronizes with Ancestry's substantial online databases.

Willing to invest a little in your genealogical research? Here are a few more of the most popular options:

Family Tree Maker:
www.ancestry.com (click on "Shop")
The most popular software and very user-friendly

The Master Genealogist (TMG):
www.whollygenes.com
An extremely flexible and feature-laden package favored by serious genealogists

For an in-depth review of many genealogical packages, see Bill Mumford's Report card: www.mumford.ca/reportcard

Web Sites

Since there are literally millions of Web sites devoted to genealogy, it's difficult to select just a few to suggest. In addition to those cited in Chapters 2 and 9, though, here are a few we use heavily and think you might want to bookmark:

www.ancestry.com
Easily the largest of all the commercial genealogy sites, this one has several subscription options—although some of the data is free. Among

their ever-expanding digitized image collections are census and immigration records as well as historical newspapers.

www.genealogy.com
This site is now owned by Ancestry.com but has not been combined with it. Strong points include its digitized and indexed 1900 U.S. census and its extensive lineage collection, World Family Tree.

www.rootsweb.com
Also part of the MyFamily.com cluster of sites, this one has many valuable location-specific ones, especially at the state and county levels.

www.familysearch.org
The Family History Library's online presence, this site enables you to search the library's entire catalog (microfilms can be ordered and then viewed at your closest Family History Center), which covers the entire globe. Other searchable collections include the Ancestral File, International Genealogical Index, and Pedigree Resource File.

www.cyndislist.com
A great resource for finding Web sites devoted to any particular aspect of genealogy that has captured your attention. Want to learn more about Canadian records, Baptists, passports, DNA, or obituaries? Cyndi has a collection of links for you.

Appendix B: DNA Testing Companies

For your convenience, we've provided contact and other basic information for most of the genetic genealogy companies (shown in alphabetical order). We say "most" because this field is evolving at a remarkable rate, and there may be new entries to the marketplace at any time. As much as we would have liked to, we have also chosen not to list prices or the number of markers examined for various tests because both are so fluid. A quick visit to the companies' Web sites will give you the latest on these details.

African Ancestry
www.africanancestry.com
5505 Connecticut Avenue, N.W.
Suite 297
Washington, D.C. 20015 U.S.A.
info@africanancestry.com
Phone: (202) 723-0900
Fax: (202) 318-0742

PRODUCTS:
PatriClan™—Y-DNA test
MatriClan™—mtDNA test

African Ancestry has a specialized database of more than 10,000 DNA samples collected from around Africa and is consequently able to provide more precise results for those of African ancestry. Approximately 70 percent of their clients find a match in their database.

DNA Heritage
www.dnaheritage.com
40 Preston Road
Weymouth, Dorset
DT3 6PZ U.K.
info@dnaheritage.com
Phone: [+]44 (0) 1305-834936
Fax: [+]44 (0) 1305-835925

PRODUCTS:
Y-DNA test

DNA Heritage's Web site provides a useful tutorial about Y-DNA. The company also hosts the open database located at www.ybase.org.

DNAPrint™ Genomics
www.dnaprint.com
900 Cocoanut Avenue
Sarasota, Florida 34236 U.S.A.
info@dnaprint.com
Phone: (941) 366-3400
Fax: (941) 952-9770

PRODUCTS:
ANCESTRYbyDNA™—autosomal test showing your ancestry in broad, geographic categories (e.g., Indo-European, Sub-Saharan African, East Asian, and Native American)

ANCESTRYbyDNA™ is also available through other companies that market it, but DNAPrint™ Genomics does the lab work for all of them. Also does paternity testing.

Family Tree DNA (Genealogy by Genetics, Ltd.)
www.familytreedna.com
1919 North Loop W.
Suite 110
Houston, Texas 77008 U.S.A.
info@familytreedna.com
Phone: (713) 868-1438
Fax: (832) 201-7147

PRODUCTS:
Y-DNA—several resolutions; refinements from earlier tests, and conversions from other companies also available; also offer a SNP haplogroup test
mtDNA—two resolutions; refinement from earlier test available
Combined Y-DNA and mtDNA—several packaged versions

Family Tree DNA offers the greatest variety of products and provides helpful project management tools. DNA samples are retained to facilitate upgrading of tests. The company also hosts the open database at www.ysearch.org and publishes a free newsletter, *Facts & Genes from Family Tree DNA*.

GEN by GEN
www.genbygen.de
Alte Dorfstr. 18
37124 Göttingen-Rosdorf
Germany
info@genbygen.de
Phone: [+]49 05502-910 93 83
Fax: [+] 49 01212-520 972 540

PRODUCTS:
Y-Genotypisierung—Y-DNA
7 Urmütter—mtDNA

Gen by Gen has English-speaking staff. You can use Internet Web page translators to read their Web site.

Gene Tree
www.genetree.com
2495 South West Temple
Salt Lake City, Utah 84115 U.S.A.
info@genetree.com
(888) 404-GENE (404-4363)
Phone: (801) 461-9757
Fax: (801) 461-9761

PRODUCTS:
Y-Chromosome STR DNA Study
mtDNA Sequence Analysis
Native American Verification—available for both Y-DNA and mtDNA

Gene Tree also offers forensics tests and a variety of close kin tests, including full- or half-siblingship, aunt or uncle, grandparentage, and first cousin.

GeoGene
www.geogene.com
Thornton House, Thornton Road
Wimbledon
SW19 4NG U.K.
Phone: [+]44 20-8405-6425
Fax: [+]44 870-7620711
E-mail via the Web site

Products:

GeoMother ™ Wallchart—mtDNA-based deep ancestry analysis, including map of migration routes
GeoFather™ Wallchart—Y-DNA-based deep ancestry analysis, including map of migration routes

Oxford Ancestors
www.oxfordancestors.com
Oxford Ancestors Ltd.
P.O. Box 288
Kidlington, Oxfordshire
OX5 1WG U.K.
enquiries@oxfordancestors.com

Products:

Y-Clan™—Y-DNA test
MatriLine™—mtDNA test showing which of 36 world clan mothers you descend from
Tribes of Britain™—Y-DNA test revealing ancestral tribal group/haplogroup (i.e., Celtic, Anglo-Saxon/Danish, Viking, and Norse Viking) for those of British origin.
Y-Clan™Plus MatriLine™—combined Y-DNA and mtDNA tests

Pioneered the "Seven Daughters of Eve" testing and provides attractive maps with test results.

Relative Genetics
www.relativegenetics.com
2495 South West Temple
Salt Lake City, Utah 84115 U.S.A.
info@relativegenetics.com
Phone: (801) 461-9760
Fax: (801) 461-9761

Products:

Paternal (Surname) Lineage Analysis—Y-DNA test
Maternal Lineage Analysis—mtDNA test
Family Genetics™ Projects—projects to properly verify the biologic connections of all participating individuals
Native American Validation Tests—available in both Y-DNA and mtDNA versions

Offers upgrades from individual tests to Family Genetics™ Projects and conversions from other testing companies. Family Genetics™ Projects are noted for their customized reports with MRCA and other charts and diagrams.

Roots for Real
www.rootsforreal.com
P.O. Box 43708
London
W14 8WG U.K.
Phone: [+]44 845-450-0180
info@rootsforreal.com

PRODUCTS:
mtDNA Tracing Service

Results include a customized map showing your maternal geographic origins.

Trace Genetics LLC
www.tracegenetics.com
1490 Drew Avenue
Suite 170
Davis, CA 95616 U.S.A.
info@tracegenetics.com
Phone: (866) 731-2312
Fax: (530) 297-5039

PRODUCTS:
Y-test (also offer SNP tests)
mtDNA Analysis
Autosomal DNA Test—shows your ancestry in broad, geographic categories (e.g., Indo-European, Sub-Saharan African, East Asian, and Native American
Native-American Ancestry Test

Trace Genetics also offers forensics tests and will undertake DNA extraction from old and unusual sources. In addition, the company has a specialized database of several thousand Native-American DNA samples and is consequently able to provide more precise results for those of Native American ancestry.

Close Kin Testing Companies:
See GeneTree and DNAPrint™ Genomics

Forensics Testing Companies:
See GeneTree and Trace Genetics LLC

DNA Preservation Kits:
Affiliated Genetics (www.affiliatedgenetics.com)
CATGee (www.catgee.com)
Code Amber (www.codeamber.org/dnakits.html)
DNAFiler (www.dnafiler.com)
Genetic Identity (www.genetic-identity.com)
For more links, see www.duerinck.com/archvdna.html

Appendix C:
Glossary

admixture—the blending of two or more geographic origins within an individual; for example, Indo-European, Sub-Saharan African, East Asian, and Native American; can be detected through BioGeographical Ancestry test

allele—one of the alternative versions of a gene or genetic marker that can exist in a particular location on a chromosome; in genetealogy, most often used to refer to the number of repeats in a given STR (e.g., 14–15–16 etc.); the variation in the number of these repeats is used to differentiate people; see also bi-allelic polymorphisms

Alu—a highly repetitive sequence found in high copy number (hundreds of thousands of times) throughout the genome, but only once on the Y chromosome; it is chopped into pieces of varying size by a restriction enzyme found in the bacterium Arthrobacter **lu**teus; see *YAP*

amino acids—building blocks of protein

ancestor of interest—a focus individual, often an alleged common ancestor; an ancestor whose Y-DNA or mtDNA descendants are sought for participation in a DNA study

ancestral chart—a form showing the direct-line ancestors of a particular individual; also called a pedigree

ancestral haplotype—the haplotype of a common ancestor, deduced from the haplotypes of two or more living descendants

Ancestry Informative Marker—a marker with alleles found in different percentages in different populations

autosomal—pertaining to a gene or genetic marker in any chromosome other than the sex chromosomes; in genetealogy, frequently used to refer to tests other than Y-DNA and mtDNA (such as the BioGeographical Ancestry test)

autosomes—in humans, the set of chromosomes labeled one to 22—all those other than the sex chromosomes that determine an individual's gender (XX or XY)

back mutation—when a mutation essentially undoes itself; (e.g., when the number of repeats in an STR decreases in one generation and then increases in a later generation (or vice versa), taking it back to the original number of repeats

bases—building blocks of DNA called adenine (A), cytosine (C), guanine (G), and thymine (T); see *nucleotide*

bi-allelic polymorphism—a polymorphism with only two possible states; often a SNP or indel, the two major kinds of UEPs

BioGeographical Ancestry test—autosomal test that reveals an individual's relative admixture; that is, geographic origins broken into broad categories by approximate percentages

broadcasting—genealogical technique in which the researcher locates people and information of interest by making it easier for others to find him; tactics include posting messages in appropriate forums on the Internet and placing queries in publications

Cambridge Reference Sequence—the complete sequence (16,569 bases long) of mitochondrial DNA; first analyzed in a woman from Cambridge, England

chromosome—the structure (containing DNA molecules packaged with some proteins) by which hereditary information is transmitted; humans have 46, which are found in the nucleus of most cells

cline—a gradual change in frequency of a haplogroup over a geographic area, with the area of highest frequency often being the point of origin; can also look at haplotype diversity within a haplogroup

coalescence—the process by which the number of our ancestors (or variations of a gene) are reduced as one goes back in time so that they ultimately converge on a single forebear shared by all (or a single version of a gene); in genetealogy, leads to conclusion of existence of Y-Adam and Mitochondrial Eve

CODIS markers—the 13 autosomal STR markers used by the FBI and included in the **CO**mbined **D**NA **I**ndex **S**ystem, a database of records with DNA profiles from criminal offenders, crime scene evidence, and missing persons and their relatives

collateral lines—branches that are descended from a common ancestor but not through the same direct line; in genetealogy, researching these lines often produces DNA-testing candidates for other lines whose DNA has daughtered or petered out

complementary pairs—the DNA molecule is made up of two long helical strands that connect together with the bases on one strand bonded to the bases on the other; adenine always fits with thymine (A to T) and cytosine always fits with guanine (C to G), so these pairings are said to be complementary

confidence interval—the calculated range with a given probability (e.g., 95 percent) that the true value of a statistic (such as a mean, proportion, or rate) is contained within the range; often used to describe a range of generations within which an MRCA lived at varying probability levels, and for tests of geographic origins

convergence—the process by which two different haplotypes mutate to become identical; causes an accidental match

cosmopolitan—used to refer to SNPs that occurred early before man migrated out of Africa, and which could consequently appear anywhere on earth; see *private* for contrast

crossing over—the process by which a stretch of DNA is swapped out and replaced by a similar stretch from a paired chromosome; see *recombination*

CRS—see *Cambridge Reference Sequence*

cytoplasm—substance inside of a cell that surrounds the nucleus; contains mitochondria, used in mtDNA testing

daughtering out—when a Y-DNA line has died out because only daughters were born; see *petering out* for reverse

deep ancestry—ancestors from more than a thousand years ago, sometimes tens of thousands of years ago; see *genealogical time frame*

descendancy chart or descendant tree—a chart showing the descendants (organized by generation) of a selected ancestor; often used in genetealogy to locate or determine relationships between testing participants

DNA—DeoxyriboNucleic Acid (dee-ox-ee-rye-boh-new-clee-ic acid); double-stranded molecule (that forms the familiar double-helix) that encodes genetic information; composed of phosphate, deoxyribose (a sugar), and the four bases (A, C, G, and T)

DNA polymerase—enzyme associated with DNA replication (It copies separated strands of DNA and uses each copy as a template for the next.) and thus critical to the duplication of genetic information; see *Polymerase Chain Reaction (PCR)*

DNA profile—a series of numbers (representing autosomal STRs) that collectively identify a person; also called DNA signature or fingerprint; see *CODIS markers*

dominant—pertains to alleles (and corresponding traits) that always manifest whenever present; see *recessive* for contrast

DYS—DNA Y Segment; a number designating a given marker on the Y chromosome (e.g., DYS19, DYS390 etc.); system established by international convention

founder effect—genetic drift that occurs when certain alleles (and corresponding traits) occur in high frequency in a particular population because all or most of them are descended from one individual or cluster of individuals who first had the allele or brought it to a new area; is most pronounced when the founding population is small and their descendants are relatively isolated from others

GEDCOM—a file format designed to allow users of different genealogical software programs to share data

gene—fundamental unit of heredity; a sequence of nucleotides on a chromosome that codes for proteins (and ultimately, some aspect of an organism's development)

genealogical time frame—an arbitrary limit, but typically 1,000 years or less; see *deep ancestry* for contrast

genetealogy—the merger of genetics and genealogy; the use of DNA to learn about one's roots

genetic distance—in genetealogy, the number of mutations required to change from one haplotype to another

genetic marker—see *marker*

genetic pedigree—Y-DNA and/or mtDNA profiles to represent each of the branches of a family tree for a selected number of generations; a traditional ancestral chart accompanied with genetic data

genome—all the genetic material in an organism or a cell (sometimes excluding mtDNA); in humans, this is the whole set of 23 chromosomes with their more than 3 billion base pairs

genotype—the actual DNA sequence (i.e., combination of alleles) for some region of interest (i.e., selected loci located on paired or unpaired chromosomes)

germ line—refers to mutations that occur when the egg and sperm are formed; such mutations are heritable and can be used to track connections between generations; see *somatic* for contrast; the causation can be spontaneous or induced

haplogroup—a large cluster of people who share the same UEP and whose ancestry converges on the person who was the founding father or founding mother; used to define genetic populations; in Y-DNA testing, mostly defined by SNPs; more loosely, a cluster of similar haplotypes

haplotype—the complete set of results from multiple sites tested on a chromosome inherited from one parent (e.g.., the Y or mtDNA); in Y-DNA testing, expressed as a series of numbers (each one representing the allele at a specific STR marker), which are compared to others' haplotypes for indications of relationship; in mtDNA testing, expressed as differences from the Cambridge Reference Sequence

haplotype diversity—the number of different haplotypes found in a given population; a high level of haplotype diversity implies a low chance of a random match between two people

homozygous—when chromosomes have two copies of the same allele at a given gene locus; occurs most often when an allele is common in the general population; contrasting word is heterozygous

indel—an insertion or deletion of one or more bases; some large indels are UEPs; e.g., YAP

independent assortment—Mendel's principle that traits are transmitted to offspring independently of one another (because the alleles for one gene segregate independently of the alleles of another gene); results in every child inheriting a different combination of alleles from the parents

induced—refers to mutations that occur when caused by outside agents, such as high-energy radiation, free radicals, viruses, and certain uncommon chemicals (e.g., mustard gas); see *spontaneous* for contrast

junk DNA—non-coding (i.e., not used for making proteins) stretches of DNA with no known function, which represent an estimated 95 percent of our DNA; because it does not affect traits or medical conditions, it acts as a silent recorder, accumulating mutations and preserving one's ancestral history; used by population geneticists to study the migrations of ancient peoples and by genealogists to learn about their origins

karyotype—a picture of all the chromosomes of an organism, often arranged in homologous pairs according to decreasing size; frequently used in genetic testing

locus—specific location on a chromosome; position where a particular marker is located

marker—a distinctive landmark that occurs in an otherwise featureless stretch of DNA; a DNA sequence with known genetic characteristics that can be tested for purposes of comparison (e.g., SNP, STR)

Maximum Likelihood Estimate (MLE)—the value associated with the highest probability; used in reporting estimates of when the MRCA of selected individuals lived and results of geographic origin tests

median—the middle value of a set of sorted numbers; 50 percent of the numbers are higher, and 50 percent are lower than the median

meiosis—the process by which a cell in a sexually reproducing organism cuts its chromosome number (of 23 pairs in humans) in half; the single chromosome sets of both the egg and the sperm are then combined to create a fertilized egg that has two of each chromosome again; a source of genetic variability

microsatellite—see *Short Tandem Repeat (STR)*

minisatellite—repetitive sequences, longer than microsatellites, with subsections that vary from person to person; one minisatellite (MSY1) is found on the Y chromosome

mitochondria—my-toe-CON-dree-uh, the plural form of mitochondrion; plentiful organelles in the cytoplasm of cells that provide energy for the cells; see *mitochondrial DNA*

mitochondrial DNA (mtDNA)—genetic material found in mitochondria; passed from mothers to their children, but only daughters are able to pass it on; useful to genealogists for learning about their maternal roots; also valuable for the identification of degraded remains

mitochondrial Eve—the most recent common female ancestor of all humans alive today

modal value (mode)—the single most common value, the most frequently observed outcome; used, for instance, to assess frequency of alleles at given markers

Most Recent Common Ancestor (MRCA)—the shared ancestor of two or more people who represents their closest (and therefore, most recent) link; for instance, the MRCA of a pair of second cousins is their mutual great-grandfather or great-grandmother

motif—a recurring theme with variations; often used to describe a common pattern (e.g., a cluster of mutations) that appears in a given haplogroup

MSY1—a minisatellite on the Y chromosome; it has a high mutation rate and has been used in some genetic studies; more difficult to analyze than STRs

mtDNA—see *mitochondrial DNA*

multi-copy markers—markers for which a testing lab cannot precisely determine which allele belongs to each marker (e.g., DYS464a, DYS464b, DYS464c, DYS464d); more difficult to interpret and compare than other markers

mutagenic—giving rise to mutations; used to refer to agents such as viruses and radiation that can cause mutations

mutation—a change in a DNA sequence (e.g., STR marker altering from 14 to 15); such heritable changes of genetic information are used by population geneticists to assess how closely related various peoples are and by genealogists to determine whether individuals share a common ancestor

mutation rate—the rate (usually measured in terms of frequency per transmission event) at which changes in genetic information occur (e.g., often reported as a generic 0.2 percent for any given STR marker, although marker-specific rates are being developed and refined); used as a basis for MRCA calculations to estimate how many generations or years ago such an individual may have lived

network diagram—an abstract diagram, using points and lines to show relationships between things or people

non-coding DNA—stretches of DNA that do not code for proteins and have no known purpose, but are useful for genealogical purposes; see *junk DNA*

non-paternity event—catch-all term for situations where the Y chromosome is unlinked from the surname; includes informal and casual adoption, infidelity, illegitimacy, etc.

Non-Recombining Y (NRY)—that portion of the Y chromosome that is passed essentially unchanged (except for occasional mutations) from father to son down through the generations; all Y-DNA tests for genealogy use markers on the NRY

nuclear DNA—DNA found in the nucleus of the cell and contributed by both parents; frequently used in close relationship testing; does not include mtDNA, which resides in the cytoplasm

nucleotide—a base plus a phosphate group; the building blocks of DNA; see *bases*

nucleus—the central region of the cell that houses the chromosomes and is separated by a membrane from the cytoplasm

outline descendant tree—variation on the standard descendancy chart that presents the same information but in a collapsed, text-only format; frequently used for reverse genealogy purposes because of its ability to capture and share many generations of details in a compact fashion

parallel mutation—when the same mutation occurs in different branches of descendants

paternity index (PI)—a measurement of the likelihood that an alleged father is the biological parent as compared with a random male; a proof of paternity measure that varies by state or judicial district

pedigree—a form showing the direct-line ancestors of a particular individual; also called an ancestral chart

petering out—when an mtDNA line has died out because only sons (who are now deceased) were born; see *daughtering out* for reverse

phenotype—observable traits of an organism (e.g., hair color); may or may not be genetically related

phylogenetic network/tree—a visual representation of the evolutionary relationships among organisms believed to have a common ancestor; applied to human populations (e.g., haplogroups), and at a more micro level, in genealogy to pictorially represent the relationships among individuals sharing a common ancestor

Polymerase Chain Reaction (PCR)—technique developed by Kary B. Mullis to mimic the replication process of the cell, allowing scientists to efficiently amplify (i.e., make millions of copies of) small, selected segments of DNA; sometimes referred to as molecular photocopying or Xeroxing; used in genetealogy to amplify samples submitted for analysis; see *DNA Polymerase*

polymorphic—coming in many forms, as in alleles at a given STR marker; considered a desirable trait for genetealogy because the different versions help distinguish people and populations

polymorphisms—(from the word roots poly = many and morph = forms) inherited differences in genetic markers among individuals and populations that play a key role in genetealogy; see *mutation* and *polymorphic*

prior knowledge—in statistics, narrows the range of theoretically possible outcomes, e.g., people with the same surname are more likely to have a recent common ancestor than people with different surnames

private—used to refer to SNPs that have occurred recently and are too limited in scope to help trace ancestral migration paths; see *cosmopolitan* for contrast

pseudoautosomal region—small portions of the X and Y chromosomes that recombine, exchanging genetic material with each other

recessive—pertains to alleles (and corresponding traits) that will only manifest when the dominant allele/trait is absent; recipient must have two copies of the recessive allele (one from each parent) for the trait to be evident; see *dominant* for contrast

recombination—swapping of DNA fragments between paired chromosomes; process by which offspring derive combinations of alleles differing from those of either parent; see *crossing over* and *meiosis*

replication slippage—mechanism to explain the increase or decrease in number of repeats for a Short Tandem Repeat marker

restriction enzyme—a bacterial enzyme for cutting DNA into smaller fragments; its action is restricted to certain DNA sequences

Restriction Fragment Length Polymorphisms (RFLP)—(pronounced riff-lips), collection of DNA fragments produced when DNA is cut (at designated sequences) with restriction enzymes; mutations in the slowly changing coding region of mtDNA result in different fragment lengths in people, and as detected by RFLP testing, constitute the formal definitions for mtDNA haplogroups

reverse genealogy—detective work associated with seeking out appropriate study candidates; involving tracing lines from the past to the present

sequence—the order of nucleotide bases in a DNA molecule (e.g., AGCTTTACGGA) that encodes for proteins; the sequence of the human genome is 3 billion DNA bases

sex chromosomes—the X and Y chromosomes; the Y is involved in sex determination and is the basis for most present genetealogical testing

Seven Daughters of Eve—Bryan Sykes's term for the founding mothers of the European haplogroups as well as the title of his best-selling book

sex-linked—refers to traits and diseases that derive from genes on the X chromosome, such as color blindness

Short Tandem Repeat (STR)—a *short* pattern (often two to five bases in length) *repeated* a number of times in a row (in *tandem*); for instance, GATAGATAGATA, three repeats of the GATA sequence; the differences in the STRs at selected markers on the Y chromosome provide a basis for comparison among individuals and populations and are used extensively for most Y-DNA genetealogical testing; also called a microsatellite

silent mutation—a mutation that is not expressed and has no effect on the phenotype of an organism; such mutations accumulate over time because they do not affect survival but can provide useful means of discriminating between people and populations

Single Nucleotide Polymorphism (SNP)—(pronounced "snip"), a small genetic change or variation that occurs within a DNA sequence when a single nucleotide, such as an A, replaces one of the other three nucleotide letters: C, G, or T; occur so infrequently that they are used to define haplogroups

Social Security Death Index—a nationwide index of most Americans (along with their birth and death dates and other details) who have died since 1962; a useful tool for reverse genealogy

somatic—refers to mutations that occur somewhere in the body and are not passed on to the next generations; the causation can be spontaneous or induced; see *germ line* for contrast

spontaneous—refers to mutations that occur when the DNA polymerase in effect makes a copying error; see *induced* for contrast

transmission event—a birth; a passing of DNA to the next generation; often used to refer to opportunities at which a mutation could have been introduced

Unique Event Polymorphism (UEP)—a class of mutations in which the mutation rate is so low that it can be considered a one-time event or essentially unique—for instance, SNPs and indels; useful for exploring deep ancestry

Y-Adam—the most recent common male ancestor of all humans alive today

YAP—Y Alu Polymorphism; Alus are found all over the genome but only once on the Y chromosome; it is an indel, one type of UEP used to define haplogroups, especially African

Y-DNA—genetic material found in the Y chromosome; passed from fathers to their sons essentially unaltered down through the generations except for occasional mutations; used for tests designed to explore one's paternal ancestry

Index

Boldface page references indicate illustrations.
Underscored references indicate boxed text or tables.

C

J

K

L

M

N